Education

Core Programming Topic at the 2012 AIChE Spring Meeting and 8th Global Congress on Process Safety

Houston, Texas, USA
1-5 April 2012

ISBN: 978-1-62276-554-6

Printed from e-media with permission by:

Curran Associates, Inc.
57 Morehouse Lane
Red Hook, NY 12571

Some format issues inherent in the e-media version may also appear in this print version.

Printed by Curran Associates, Inc. (2012)

For permission requests, please contact AIChE
at the address below.

AIChE
3 Park Avenue
New York, NY 10016-5991

Phone: (203) 702-7660
Fax: (203) 775-5177

www.aiche.org

Additional copies of this publication are available from:

Curran Associates, Inc.
57 Morehouse Lane
Red Hook, NY 12571 USA
Phone: 845-758-0400
Fax: 845-758-2634
Email: curran@proceedings.com
Web: www.proceedings.com

TABLE OF CONTENTS

(4a) Equations of Change: How Patent Law Has Changed, and What Changes Are to Come

Jeffrey L. Wendt, The Wendt Firm PC, The Woodlands, TX

Patent law, particularly in the United States, has undergone many changes in the last 20 years, both in the courts and in the patent statute. My focus will be on what those changes have been, what changes are likely to come, and how these changes impact practicing chemical engineers. For example, 20 years ago the so-called "doctrine of equivalents" was widely used to ensnare infringers of patents. Now, the doctrine is almost an afterthought. My thoughts will touch on whether this is good or bad. Another change has been the term of patents. Before 1995, an issued patent had a term of 17 years from issue, even if the patent application had been filed many years before. Now, patents are granted for 20 years from the earliest filing date claimed as priority. Finally, I will touch on how a change to a "first to file" patent regime will make life different for engineer/inventors than the present "first to invent" system, particularly for small entities.

(4b) What Is Prior Art in the "First-to-File" Patent System Under the America Invents Act of 2011

Mark Rygiel, Sterne Kessler, Washington, DC and Anbar Khal, Sterne, Kessler, Goldstein & Fox PLLC, Washington, DC

On September 16, 2011, the Leahy-Smith America Invents Act was signed into law by President Obama. This law brings sweeping change to the current U.S. patent system, including transforming the current first-to-invent system into a first-inventor-to-file system. The first-to-file provisions of the Act take effect on March 16, 2013. These provisions substantially overhaul what acts are considered to be "prior art" and what acts are excepted. This presentation will provide an in-depth review of the new scope of prior art under the America Invents Act of 2011, focusing on what forms of disclosure prior to the effective filing date of a patent application will prevent the applicant from obtaining a patent in the United States. Case studies will be presented to illustrate the new prior art provisions, their exceptions, and how the new provisions compare to the existing prior art standards. The presentation will touch on how these new prior art provisions compare to the prior art rules of other developed countries having first-to-file systems.

Avoiding Common Mistakes in Licensing Agreements and Other Technology Agreements

By: Anthony P. Venturino, Esq. , Partner of Novak, Druce + Quigg, LLP Washington, DC

AIChE Spring Meeting - April 2, 2012 Houston, TX

The views presented here are those of the author and not attributable to his firm or its clients.

I. Technology agreements Can't live with them, can't live without them

Whether you consider technology agreements a help or a hindrance we need them to control the flow of technology. They set the ground rules so everyone involved knows how to handle technology being transferred and/or invented so there are no unwelcome surprises.

Thus, chemical engineers and managers need to know enough to know when they need an agreement or, how to perform if they are operating under an agreement.

In the course of performing their jobs chemical engineers and managers encounter many kinds of agreements having technology or other forms of intellectual property as a critical component. A few such agreements are as follows:

- Confidentiality Agreements
- Vendor Supplier Agreements
- Joint Development/Commercialization Agreements
- Exclusive Licenses / Non-exclusive Licenses
- Co-promotion Agreements
- Employee Agreements
- Settlement Agreements
- Joint Venture Agreements
- Contract R&D Agreements
- Consulting Agreements

II. Things can go wrong before, during and after the agreement

Here are some examples of where things can go wrong:

- Evaluation of the technology
- Selection of partners
- Conduct of the negotiations
- The agreement itself
- Implementing and servicing the agreement

Classic Licensee mistakes include:

- Royalties paid when licensed patent had expired
- Payments made after royalty terms had been satisfied
- Payments made under a trademark license when the trademark is not used

- Patents covered in a license no longer cover products that are currently made, yet royalty is paid against the patent coverage
- Royalties paid at the wrong rate, i.e. one higher or lower than provided Source:

Licensing – A Strategy for Profits, Edward P. White (1990)

Licensors also make mistakes. Here are some classics:

- Failure to take action if royalty payments cease when they should not, e.g., prior to expiration of the patents.
- Failure to monitor royalty payments, e.g. sudden reductions or failure to meet minimum royalties that could trigger a change from exclusive to nonexclusive or termination of the licenses.
- Failure to stay in technical contact with the licensee, with the result of licensee not interested in paying royalties, in maintaining quality, or keeping good records.

Licensing – A Strategy for Profits, Edward P. White (1990)

Before doing the deal.

- Know your business goals, strengths and weaknesses
- Know the available protection for the technology
- Have a realistic idea of the value of the technology
- Be open to creative technology agreements
- Do not let paranoia get in the way of a good deal
- Watch out for foreseeable legal and technical problems
- Admit when you are in over your head and get help
- Admit when you are in over your head and get help from a lawyer or other licensing professional (did I just say that?)

III. Avoiding Common Traps

A. Define ownership and control of joint inventions of improvements

Jointly developed invention often occur when smart people from different companies work together, for example while performing a joint R & D agreement or a variety of other agreements between companies. Joint inventions are jointly owned by US law unless there is a contract stating otherwise. This can cause problems.

If possible, have one party own the invention and, if necessary, grant the other party an exclusive field of use license. The following clause does this:

> In the event a joint invention arises during performance of this Agreement Company A owns the joint invention and controls patent prosecution and Company B gets an exclusive field of use license to the joint invention in the field of ______________.

B. Maintain good records of what the agreement covers or when it expires

The company, its legal department or a law firm should docket the agreement in a database. The docketing record should include: the type of agreement, the parties, the technology covered by the agreement, the agreement deadlines, the expiration date, the contact person and backup contact person, and have a miscellaneous portion for anything unusual.

Preferably an institutional person, for example, a docket clerk or contracts administrator, should be responsible for tracking notices in the docketing system and then relay the information to the business people. The more automatic you can make the timely conveying of such information, the better.

C. Do not accidentally create an agreement

The first time you negotiate with a party, especially if the other party sees this agreement as a one shot deal, be careful not to make records during the negotiations which unexpectedly bind you. Do not let an agreement to agree inadvertently become an agreement.

Summaries of the negotiations signed by both parties can lead to binding agreements even when a clause that might be important to you is missing. Let's take a hypothetical scenario where two parties bring their business people and lawyers to a meeting to negotiate a potential IP license. Mid-afternoon the attorney for the IP owner (potential licensor) had to leave to catch a plane. The potential licensor's business person (let's call him "Bill") had a later plane. The potential licensee kindly offered to take Bill to the airport later so he would not have to waste time in a boring airport. While Bill was by himself with the potential licensee's attorney and business people one of the business people suggested they draft and sign a summary of the day's discussions. They did this and then Bill caught his plane. The next day Bill's company was asked to sign an Assignment document to reflect the assignment they agreed to yesterday. Bill's response was "What assignment?" After a lawsuit the parties settled. The moral of the story is train the business people not to sign anything.

A deal can also inadvertently arise if the parties act like there is an agreement can create an agreement. This sometimes arises where the parties are in a rush to do joint development. While the deal is only a draft they start working together. Sometimes the unsigned draft becomes the *de facto* agreement.

The opposite can also happen. Sometimes you think you have an agreement but in reality you do not, or at least not all the clauses apply.

Take the case of *DataTreasury v. Wells Fargo,* 522 F.3d 1368 (Fed. Cir. 2008). In 2003 WMR e-Pin licensed a patent to Wells Fargo Services Corp. a subsidiary of Wells Fargo & Co. The license included an arbitration clause. In 2006 WMR e-Pin assigned 4 patents to DataTreasury. At least one of the 4 patents was a patent of the license. Afterwards DataTreasury sued Wells Fargo & Co. for patent infringement in court. Wells Fargo & Co. moved to dismiss or stay pending arbitration. Under 5th Circuit law, as applied by the

Federal Circuit, "to be enforceable, an arbitration clause must be in writing and signed by the party invoking it. *DataTreasury* at 1373, quoting *Westmorland v. Sadoux*, 299 F.3d 462, 465 (5th Cir. 2002). Neither Wells Fargo & Co. nor DataTreasury signed the original patent license. Thus, the court held these parties were not bound by the arbitration clause.

D. Remember important clauses

Clause checklists are important tools when reviewing a license to see if all the necessary clauses are present. However, use the right checklist for the type of agreement and the type of IP. Do not assume the clauses in your standard patent license belong in your new trademark license and *vice versa*. Unlike patents which protect the patent holder, trademarks protect the public by assuring the public a good labeled with a particular trademark will have the characteristics they expect. For example, a PIRELLI brand tire has certain handling properties. Trademark licenses without quality safeguards are termed naked licenses and invalidate the trademark rights.

A short checklist is provided in TABLE 1:

TABLE 1
Introductory Clauses
Definitions
Grant of Rights
Consideration/Compensation
Diligence
Product Quality Control
Advertising
3rd Party Infringement Suits
3rd Party Product Liability Suit
Licensor/ Licensee Suits
Termination
Continuing Obligations

A detailed set of clause checklists for a licensing agreement is available upon request from the author.

E. Maintain control of the technology after making the deal

Proprietary technology is a competitive advantage. If the licensor loses control of the technology he loses control of the competitive advantage. Thus, the licensor should determine the level of control he is willing to lose. Maybe the licensor thinks he can permit the licensee to sublicense because he trusts the judgment of licensee's president. What if the president leaves? Thus, the agreement preferably precludes sublicensing or assigning to competitors.

Another common way to lose control is for the licensee to be bought by the licensor's competitor. The Agreement should specify it may be assigned by ONE of the following:

1. Anyone;
2. Goes with the business;
3. Goes with the company;
4. Licensor maintains right of first refusal;
5. Licensor maintains right of first refusal which will not be unreasonably withheld;
6. No one.

F. Promote good communication after making the deal

Problems arise from misunderstandings, lack of thoroughness, carelessness, or failure to administer the agreement at all, i.e., no one in authority is responsible for administration.

The lawyer can structure the agreement to include mechanisms for communication between the licensor and licensee so little problems and misunderstandings do not become big ones. You cannot force the business people to communicate but the mechanisms provide fertile soil for communication. For example, the agreement should include a contact person and back-up contact person. The agreement can also mandate times for periodic meetings.

G. You need the equivalent of a pre-nuptial agreement

Agreements, like romance, can end badly. As with a newlywed couple, the enthusiastic partners to a new business relationship think nothing will go wrong. They can and they do.

Include termination provisions in the agreement to give each party a reasonable way to terminate the agreement. This includes planning for unexpected events, for example, bankruptcy or purchase of the other party. This also includes providing ways to terminate or renegotiate if the economics of the deal no longer make the current agreement viable.

IV. Organize A Licensing Department – A final word

Some companies only occasionally do a license or may be at the beginning of setting up a license project group. In these cases, the company might ask his or her lawyer for advice on how to organize the licensing project group.

The license project group will need the support of a top executive. Without this the license project group will generally not get the resources it needs.

Recognize the different functions to be performed. Then, divide or share these functions depending on available resources, business people, consultants or lawyers:

- Profit Center Group - optimizes the licensing in and licensing out to make money for the company
- Support Group - coordinates technical, legal and accounting aspects of tech transfer
- Administrative Group -handles accounting, record keeping and day to day details of maintaining the license.

V. Conclusion

Technology agreements help people work together while managing the intellectual property of their efforts. These agreements let everyone involved know their rights and responsibilities. Knowing your goal and having the appropriate agreement to help get to the goal and understanding the agreement can avoid nasty surprises. If you know your goals, technology and agreements you will be in a better position to turn that technology into profit.

MISAPPROPRIATION OF TRADE SECRETS BY DEPARTING EMPLOYEES OR OTHER AGENTS

Teresa J. Lechner-Fish
Chris P. Perque
Gardere Wynne Sewell, LLP
Houston, TX
www.gardere.com
Telephone 713.276.5500
Email tlechnerfish@gardere.com
Email cperque@gardere.com

I. INTRODUCTION

This paper discusses a common business tort: misappropriation of trade secrets by departing employees or other agents. A trade secret may be any information that can be used to economic advantage. However, to be entitled to trade secret protection, such information must be maintained as a secret. Valuable proprietary information can be lost when employees or other agents depart to a competitor. Accordingly, protection of such proprietary information must begin at the onset of the employee's or agent's employment and continue throughout the employment. Protection and enforcement of trade secrets is discussed for various jurisdictions with an emphasis on recent cases in California and Texas.

II. MISAPPROPRIATION OF TRADE SECRETS BY DEPARTING EMPLOYEES OR OTHER AGENTS

Most states have adopted some form of the Uniform Trade Secret Act (UTSA), however, Massachusetts, New Jersey, New York and Texas have not adopted the Act. A few states including Texas and New York refer to the Restatement of Torts.

Preliminary injunctions may be obtained in special circumstances to prevent irreparable harm to the trade secret owner. Examples of recent preliminary junction cases in California and Texas are discussed below.

A. California

Like most states, California has adopted the Uniform Trade Secrets Act. *See* California Uniform Trade Secrets Act, CAL. CIV. CODE §§ 3426 *et seq.* Under the California Uniform Trade Secrets Act,

> (b) "Misappropriation" means: (1) Acquisition of a trade secret of another by a person who knows or has reason to know that the trade secret was acquired by improper means; or (2) Disclosure or use of a trade secret of another without express or implied consent by a person who: (A) Used improper means to acquire knowledge of the trade secret; or (B) At the time of disclosure or use, knew or had reason to know that his or her knowledge of the trade secret was: (i) Derived from or through a person who had utilized improper means to acquire it; (ii) Acquired under circumstances giving rise to a duty to maintain its secrecy or limit its use; or (iii) Derived from or through a person who owed a

duty to the person seeking relief to maintain its secrecy or limit its use; or (C) Before a material change of his or her position, knew or had reason to know that it was a trade secret and that knowledge of it had been acquired by accident or mistake.

'Trade secret' means information, including a formula, pattern, compilation, program, device, method, technique, or process, that:

(1) Derives independent economic value, actual or potential, from not being generally known to the public or to other persons who can obtain economic value from its disclosure or use; and

(2) Is the subject of efforts that are reasonable under the circumstances to maintain its secrecy.

See § 3426.1(b) and (d); *see also ReadyLink Healthcare v. Cotton*, 24 Cal. Rptr.3d 720, 727-28 (Cal.Ct.App. 2005).

In *ReadyLink*, the plaintiff was a licensed healthcare service provider with a nurse staffing division, which provided healthcare facilities and professionals with nurses. *See ReadyLink*, 24 Cal.Rptr.3d at 723. The plaintiff recruited, interviewed and hired nurses as its own employees and assigned them to its customer healthcare providers who requested nurses. *Id.* The plaintiff invested time, money and effort in developing its database of nurses and healthcare provider customers. *Id.* at 723-24. The database was used to recruit nurses and to service its clientele. *Id.* at 723.

The plaintiff also employed a sales force to recruit nurses, secure new business and service existing customers. *See ReadyLink*, 24 Cal.Rptr.3d at 724. Plaintiff hired the defendant, and, about 6 months later, it moved him into a nurse recruiting agent position. *Id.* The defendant signed 1) upon his employment, an employment agreement and an employee confidentiality, non-disclosure and non-circumvention agreement; and 2) upon his move into the nurse recruiting agent position, an agent agreement and another proprietary and confidential information, non-disclosure and non-circumvention agreement. *Id.*

About 14 months later, the plaintiff fired defendant for stealing company records containing proprietary and confidential information. *See ReadyLink*, 24 Cal.Rptr.3d at 724. While employed with the plaintiff, the defendant was caught on videotape entering company private offices at night and copying payroll practices, confidential documents and notes. *Id.* The defendant authorized a search of his residence and storage locker. *Id.* During the search, the police recovered 1-2 boxes of plaintiff's stolen documents. *Id.* Although the police investigated the theft, the defendant was not criminally prosecuted. *Id.*

Upon termination of his employment, the defendant signed a declaration acknowledging 1) his misappropriation and improper use of plaintiff's proprietary and confidential information, 2) his attempt to form his own company in competition with plaintiff, and 3) his improper solicitation of plaintiff's employees to staff his company and investors to finance his company. *See ReadyLink*, 24 Cal.Rptr.3d at 724. A mere two weeks after his termination, the defendant began working at Registry Network, a competitor, despite agreeing in his declaration to not be employed at any other healthcare staffing company. *Id.* The plaintiff brought suit against defendant and Registry Network for, *inter alia*, violating the California Uniform Trade Secrets Act, alleging defendant stole confidential information regarding the company's finances, clients, employee contacts, payroll practices and business methods.[1] *Id.* at 724-25. The plaintiff requested injunctive relief, as well as damages, prohibiting the defendant from disclosing the plaintiff's proprietary information to Registry Network and from soliciting plaintiff's employees and customers. *Id.* at 725. The trial court granted a preliminary injunction. *Id.*

Upon interlocutory appeal, the defendant contended that the trial court abused its discretion in granting the preliminary injunction. *See ReadyLink*, 24 Cal.Rptr.3d at 722 & 726. The Court of Appeals affirmed the preliminary injunction order, as modified. *Id.* at 723 & 730. The Court stated:

> ...***in deciding whether to issue a preliminary injunction, a trial court weighs two interrelated factors: the likelihood the moving party ultimately will prevail on the merits, and the relative interim harm to the parties from the issuance or nonissuance of the injunction.***

Id. at 726-27 (quoting *Whyte v. Schlage Lock Co.*, 101 Cal.App.4th 1443, 1449-50, 125 Cal.Rptr.2d 277 (Cal.Ct.App. 2002)) (emphasis added). The Court found ample evidence supporting the trial court's finding that plaintiff would likely prevail on its misappropriation of trade secrets claim:

(1) in defendant's declaration upon termination, the defendant admitted that he misappropriated proprietary and confidential information from plaintiff, and that a disclosure of such information violated local, state, and federal law, and he also admitted that, as a consequence of misappropriating such information, it would not be possible for him to engage in the healthcare staffing business without indirectly or inadvertently using or disclosing such information;

(2) in plaintiff's declarations supporting the injunction, plaintiff presented overwhelming evidence that the defendant intended to start his own nurse staffing business in competition with plaintiff using plaintiff's trade secret information, and that he solicited plaintiff's employees and nurses using plaintiff's proprietary and confidential information such as plaintiff's databases containing lists of nurses, employees, and healthcare facility customers, compilations of compensation, employment preferences, contact information, nurse applications and tests, and plaintiff's per diem program;

(3) the supporting declarations sufficiently establish that plaintiff's trade secret information has potential economic value because the plaintiff went to great expense to compile the data, and that the information would enable a competitor to recruit nurses and employees away from the plaintiff; and

(4) the defendant entered into two employment agreements in which he agreed not to use or disclose plaintiff's proprietary and confidential information.

Id. at 728. Therefore, "it is reasonably likely that the plaintiff will prevail on its misappropriation claims." *Id.* at 730.

After finding that the plaintiff would likely prevail on its trade secrets misappropriation claims, the Court reviewed the relative interim harm to the parties from the issuance or nonissuance of the

---

[1] After plaintiff notified Registry Network of the defendant's theft of its confidential and proprietary information, Registry Network terminated the defendant's employment. *Id.* at 725.

injunction, and found ample evidence that defendant posed a threat to plaintiff's business operations at the time of the injunction hearing, including:

(1) the defendant intended to create his own competing business, and he had taken significant steps to carry out his business plan, even though he had agreed not to commit such illicit acts;

(2) shortly before his termination, the defendant was caught on videotape stealing plaintiff's documents, and a search of his home and computer revealed he had misappropriated plaintiff's documentation and information;

(3) the defendant acknowledged he misappropriated proprietary and confidential information, and he solicited plaintiff's employees and nurses using the information even though he had agreed not to use or disclose the information; and

(4) right after leaving plaintiff, the defendant secured a job at a competitor and continued soliciting plaintiff's employees and customers using plaintiff's proprietary information, while simultaneously pursuing his plan of creating his own healthcare staffing business.

Id. at 732. Because the plaintiff also satisfied the irreparable harm requirement, the Court affirmed the preliminary injunction order, as modified. *Id.* at 734.

California companies should guard the secrecy of their proprietary and confidential information by marking confidential documents as "confidential to company" and storing original documents in a vault, limiting access to information on its internal computer system to persons with appropriate system identification and passwords, requiring persons entering the company premises to identify themselves and to wear a badge, checking materials in and out of the premises, and using confidentiality agreements with employees upon employment and, again, upon reassignment to another position.

For particularly sensitive information, companies should use video cameras to record activity in areas where confidential information is kept, or where copiers are located. If an employee is caught stealing such information, the employer should require the employee to sign a declaration admitting to wrongdoing and agreeing to non-compete provisions upon termination.

California employees should avoid accessing their employer's proprietary and confidential information without its express or implied consent to do so.

B. Texas

Texas—like New York—has not adopted the UTSA—it applies common law referring to the Restatement (First) of Torts. To establish misappropriation of trade secrets, a plaintiff must show:

(1) existence of trade secret;

(2) trade secret was acquired through a breach of a confidential relationship or was discovered by improper means; and

(3) defendant used the trade secret without the plaintiff's authorization.

See IAC, Ltd. v. Bell Helicopter Textron, Inc., 160 S.W.3d 191, 197 (Tex.App. –Fort Worth 2005). Texas courts have adopted the definition of a trade secret from the commentary to Section 757 of the Restatement of Torts:

> Any formula, pattern, device or compilation of information which is used in one's business and presents an opportunity to obtain an advantage over competitors who do not know or use it.

See IAC, 160 S.W.3d at 197 (quoting *In re Bass*, 113 S.W.3d 735, 739 (Tex. 2003)); *see also* RESTATEMENT (FIRST) OF TORTS § 757 cmt. b (1939). To determine whether information is entitled to trade secret protection, courts consider the *Bass* factors:

(1) the extent to which the information is known outside the holder's business;

(2) the extent to which it is known by employees and others involved in the holder's business;

(3) the extent of the measures taken by the holder to guard the secrecy of the information;

(4) the value of the information to the holder and its competitors;

(5) the amount of effort or money expended by the holder in developing the information; and

(6) the ease or difficulty with which the information could be properly acquired or duplicated by others.

See IAC, 160 S.W.3d at 197-98 (citing *In re Bass*, 113 S.W.3d at 739); *see also* RESTATEMENT (FIRST) OF TORTS § 757, cmt. b.

In *IAC*, plaintiff was in the helicopter design and manufacture business. *See IAC*, 160 S.W.3d at 196. The defendants were in the helicopter maintenance and service business, and were trying to be in the helicopter replacement blade business. *Id.* In 2001, Jim Van Horn formed Van Horn Aviation, LLC to develop a 206B replacement blade. *Id.* Randy Stevens, principal of IAC, Ltd., agreed to manufacture and market the replacement blade. *Id.* At the time, IAC, Ltd. was a customer service facility (CSF) for the plaintiff. *Id.* As a CSF, IAC, Ltd. received proprietary data relating to the plaintiff's 206B rotor blade. *Id.* Stevens agreed to use the proprietary data disclosed by plaintiff solely for the purpose of repairing plaintiff's helicopters. *Id.*

Knowing that plaintiff's OH-58 blade is similar to its 206B blade, Van Horn purchased a Depot Maintenance Work Report (DMWR) for plaintiff's OH-58 helicopter from Newport Aeronautical Sales. *Id.* Van Horn used the OH-58/206B contour and twist data and 206B sealants and adhesives for the replacement blade. *Id.* at 197. When the plaintiff learned that defendants were making a replacement blade, it sued for, *inter alia*, misappropriation of trade secrets, and theft of trade secrets. *Id.* The trial court granted a temporary restraining order, and, about two months later, a temporary injunction. *Id.*

Upon interlocutory appeal, the defends contended that, *inter alia*, the trial court erred in granting the temporary injunction. *See IAC*, 160 S.W.3d at 197. The Texas Court of Appeals affirmed the temporary injunction. *Id.* at 203. The Court of Appeals stated:

> ***To be entitled to a temporary injunction, the applicant must plead a cause of action and show a probable right to recover on that cause of action and a probable, imminent, and irreparable injury in the interim***...

Id. at 197 (internal citations omitted) (emphasis added). The Court found that the plaintiff's evidence, when viewed in light of the *Bass* factors, supported the trial court's finding that the 206B data that plaintiff disclosed to IAC, Ltd. and the DMWR obtained by Van Horn are entitled to trade secret protection at least until a trial on the merits. *Id.* at 198. With respect to the 206B data, the plaintiff produced evidence showing the measures it takes to guard the secrecy of the data, the amount of effort plaintiff expended in developing the data, the data's value, and the difficulty with which the data could be duplicated by others:

(1) plaintiff guards the secrecy of its data by storing the originals of its drawings and specifications in a vault, posting security guards at its plants, requiring persons entering the plant to identify themselves and wear identification badges, checking material going in and out of the plant, limiting access to data on its internal computer system to persons with appropriate system identification and passwords, and using confidentiality agreements with its vendors to whom it supplies proprietary data;

(2) one of plaintiff's engineers testified that thirty to forty thousand hours were required to develop the design of the blade, and another thirty to forty thousand hours were needed to develop techniques for its production;

(3) one of plaintiff's sales and marketing employees estimated that the replacement blade market was worth approximately twelve million dollars a year; and

(4) expert testimony showed that it would not be possible to duplicate the blade specifications with the necessary degree of precision through reverse engineering.

Id. at 198. With respect to the DMWR, the plaintiff produced evidence of the measures it takes to protect the secrecy of the data contained in the DMWR and the extent to which the information is known outside of plaintiff's business:

(1) plaintiff showed that the DMWR was created using plaintiff's drawings that would have been marked "confidential to Bell" and that the DMWR stated on its face that it was not available through Attorney General publication centers; and

(2) plaintiff's evidence also revealed that the drawings used to create the DMWR were not available through a Freedom of Information Act (FOIA) request, and that Newport Aeronautical, Van Horn's alleged source of the DMWR, provided the DMWR only on a "need to know basis" and did not provide electronic access to it.

Id.

The Court also found that the plaintiff's evidence supported the trial court's finding that the defendant acquired the 206B data and DMWR through a breach of a confidential relationship or discovered it by improper means. *See IAC*, 160 S.W.3d at 199. With respect to acquiring the information through a breach of a confidential relationship, the plaintiff produced evidence of IAC, Ltd.'s probable breach of the confidential relationship with the plaintiff to obtain the 206B data:

(1) several witnesses testified, and plaintiff's Customer Service Facility Agreement showed, that plaintiff disclosed 206B data to IAC, Ltd. so that IAC, Ltd. could repair plaintiff's helicopters;

(2) IAC, Ltd. had an obligation to keep the plaintiff's data confidential pursuant to a confidentiality agreement;

(3) Stevens, the president of IAC, Ltd., was a partner with Van Horn in the defendant's 206B replacement blade project;

(4) IAC, Ltd. kept the 206B data in its library, and Van Horn had access to the data; and

(5) the contour and twist data for the 206B blade is identical to that for the replacement blade.

Id. With respect to acquiring the information through improper means, the plaintiff also produced evidence of Van Horn's likely improper means to obtain the DMWR:

(1) the DMWR was not available through Attorney General publication centers;

(2) the drawings upon which the DMWR was based were not available through a FOIA request; and

(3) Newport Aeronautics provided the DMWR only on a "need to know" basis.

Id.

In addition, the plaintiff produced some evidence that defendant used the 206B data and DMWR without the plaintiff's authorization:

(1) the defendants used contour and twist data contained in both the 206B data disclosed to IAC, Ltd. and the DMWR and that they used primer and adhesive specifications present in the 206B data;

(2) Van Horn created his certification plan to submit to the Federal Aviation Administration for the 206B project on the same day he purchased the DMWR;

(3) IAC, Ltd.'s was expressly prohibited from using the 206B data pursuant to a confidentiality agreement, and Van Horn was not authorized to use the data in the DMWR.

Id. at 199-200. "Because [plaintiff] presented some evidence to support each element of its trade secret misappropriation claim, it has shown a probable right to recover under this cause of action." *Id.* at 200.

After finding that the plaintiff would probably recover on its trade secrets misappropriation claim, the Court reviewed whether the plaintiff met its burden of showing that a probable, imminent, and irreparable injury would result before trial in the absence of a temporary injunction. *See IAC*, 160 S.W.3d at 200. Plaintiff's evidence showed that the defendants had possession of plaintiff's data entitled to trade secret protection and they were actively using that information to compete with plaintiff in the replacement blade market. *Id.* Further, plaintiff's evidence also showed that defendant's replacement blades were going to be available for shipment around June 15, 2004 and would be priced lower than plaintiff's blade:

> [Plaintiff's] Executive Director of Integrated Product Solutions testified that the creation of a lower-priced 206B replacement blade market would undermine the goodwill that Bell has with its clients, leading to an incalculable loss of business goodwill.

Id. at 200. Because the plaintiff also satisfied the imminent and irreparable injury requirement, the Court held that the trial court did not abuse its discretion by granting the temporary injunction. *Id.*

Texas companies should guard the secrecy of their proprietary and confidential information by marking confidential documents as "confidential to company" and storing original documents in a vault, limiting access to information on its internal computer system to persons with appropriate system identification and passwords, requiring persons entering the company premises to identify themselves and to wear a badge, checking materials in and out of the premises, and using confidentiality agreements with employees and vendors. For particularly sensitive information, companies should use video cameras to record activity in areas where confidential information is kept, or where copiers are located.

Texas vendors, that are subject to a confidential agreement, should guard the secrecy of the other party's confidential information as discussed above.

Chris Perque is a partner with Gardere Wynne Sewell LLP in Houston, Texas specializing in intellectual property and commercial litigation.

Teresa Lechner-Fish is an associate with Gardere Wynne Sewell LLP in Houston, Texas specializing in intellectual property and commercial litigation.

The views expressed herein are solely those of the authors and do not necessarily reflect the views of Gardere Wynne Sewell LLP or its clients.

(36b) Environmental Liability: Regulation, Compliance and Litigation Risk

Mary Ellen Ternes, McAfee & Taft, Oklahoma City, OK

Compliance with environmental regulations, while necessary and helpful in minimizing litigation risk, does not completely insulate industry from litigation. This presentation will review potential pathways towards unwanted administrative and judicial entanglements regarding which even the most compliant companies should be aware, including enforcement disputes, citizen suit actions and common law claims.

Title: Propellant Systems for Household Spray Products
Submitter's E-mail Address: cjabraham1@aol.com
Preferred Presentation Format: Oral
Keywords: Hazards, Standards and Total Release Foggers
Session Selection: Litigation Perspectives for Chemical Engineers
AICHE – Houston, TX April 1-5- 2012

PROPELLANT SYSTEMS FOR HOUSEHOLD SPRAY PRODUCTS

C. J. Abraham, Scientific Advisory Services, Ltd, Great Neck, NY

Abstract

The problem delineated in this paper involves consumers throughout the United States who use and are exposed to bug bombs or bug foggers (also known as total release foggers orTRF's). They are the most dangerous aerosol products on the market. We will focus on the defective design of the formulation, the state of technology available to eliminate the risks, the defects in the testing of the product, the false information that is submitted to the EPA and the lies stated in the application process and registration of pesticide products.

The most popular propellant system for total release foggers for indoor use consists of propane, butane and isobutene. Even though the propellent system consists of up to 35% of the ingredients, applications to the EPA for approval of insecticides designate the propellant system as inert. This designation is incorrect and fraudulent. In addition to the products being defectively designed, they contain a hidden hazard resulting in fires, explosions and traumatic burn injuries.

There are several resolutions which can make the product either completely safe or minimize the risk of consumers exposing themselves to explosions and toxic vapors. This involves alternative safer propellant systems and significant changes in the testing of the product, proper reporting of information in the MSDS's, proper testing and significant changes in the warnings.

Introduction

Over the last forty years this author has been involved with dozens of fires and explosions involving a popular propellant used in a number of products including total release foggers also known as TRF's. The propellant is composed of propane, butane and isobutene. The insecticide is purchase by the lower income segment of our population.

The last case that I was involved in was typical of what could happen with the TRF's. Two women, aged 25, and 26, were severely burned when the explosive and extremely flammable propellant from a popular insecticide product came in contact with an ignition source in the home of the mother of one of the women about seven hours after her mother released a number of canisters in her home. The women were burned over 70% to 90% of their bodies. Both women sued the manufacturer, retailer and distributor of the TRF insecticide.

Both women lived in a rural area of, Nebraska and the mother lived in Iowa. Prior to arriving at the mother's home, they all met at a store in between. The women were not informed of the canisters,. The women proceeded to the mother's residence to pick up two dogs.

Once the women arrived at the residence, they went in through the front door. The house needed cleaning up so both women decided to surprise the mother with a clean house. when she arrived sometime later.

Prior to the explosion, neither women detected any odor in the house other than the odor from the animals. After some time, both of them became exhausted and lay down on the couch. They woke up to the barking of dogs. They rose up to see what the dogs were barking about. When they were approximately three to four feet from the rear of the house, they heard the furnace "click" on. An explosion and flashover followed and both of them experienced the results and extreme heat of the flashover. As soon as they heard the explosion, one of them noticed that she was on fire. They went out the back door dropped and rolled on the ground.

It is important to note that neither of the women detected any odor of a liquid, chemical or vapor. That includes the gasoline which was located in a plastic container placed on the dryer in the washroom. There was also a very small amount of the gasoline remaining on the floor in the bathroom from a prior spill. Neither women used the hot water or adjust the thermostat. In addition, the lights in the living room, kitchen and bathroom were on when they arrived.

The mother never told her daughter or her freind that she set-off eight (8)TRF's throughout the house prior to her leaving. The mother's grandchildren were expecting to visit her the following week and they were allergic to spiders. The mother had purchased the TRF's at a "Dollar" store. The instructions suggest that the consumer can use one fogger per room. The label states that the four-can package is able to treat 2,000 cubic feet of unobstructed area. The subject house is approximately 1,200 square feet. There are no statements indicating what would happen if extra TRF's were used throughout the house.

The explosion and fire that occurred destroyed the home. Initially, the fire department did not perform a complete analysis with reference to the cause and origin. For example, if gasoline vapor was the cause of the fire, why wasn't the container of gasoline on top of the dryer involved with the fire? The fact is the gasoline container was still intact after the house was destroyed. The Fire Department concluded that the cause of the fire was the propellant from the TRF coming in contact with the ignition source of the furnace.

Background- The Product

Total release foggers can be purchased in many grocery, hardware stores and chain stores They are relatively inexpensive and have been used for several decades in lieu of using a professional exterminator. In order to properly use those products, one must disconnect all the electrical appliances. This includes refrigerators, freezers, lights and anything else that uses electricity. In addition, the furnace, gas stoves, hot-water heaters and all other sources of ignition must be turned off. It should be noted that all of the above is the responsibility of laypeople with limited or no education in this area. The total release foggers are turned on throughout the apartment or house and the area is left completely closed. The owners return a number of hours later with the expectation that the cockroaches, spiders and other insects are killed. The manufacturer's instructions state and assume that after a number of hours the previous explosive environment is safe to enter and that everything that was turned off previously can now be turned on again. Not one manufacturer informs the consumer that an explosion-proof ventilation system should be used to make sure that most, if not all, of the concentration of the explosive propellant is dissipated. The manufacturers also know that the average layperson does not have any type of gas-detection equipment, so he would not be equipped to even measure or detect explosive gases. The gases are invisible and are not odorized. Therefore, the consumer is exposed to a hidden danger that is known to the manufactures but not known to the consumers.

The Consumer Product Safety Commission (CPSC) has reported accidents involving similar propellant systems from the early 70s to the present time. One of the many cases that the undersigned has consulted on was reported in the United States District Court Decision: Allison Kowak, et al Plaintiffs vs. Fabrege U.S.A. and Precision Valve Corp., Defendants. Civ. No. CV 90-1919. Nov. 13, 1992. In addition, a demonstration was performed by the undersigned on CBS TV shortly after the appellate decision was published (812F.Supp. .942).

For many decades, thousands of explosions have occurred where total release foggers (TRF's) were used in apartments and homes throughout the United States. These explosions resulted in fires, destruction of property, burn injuries and death. They occurred mostly in apartments of the segment of our population that is made up of lower-income families that cannot afford professional exterminators.

In most instances, the TRF is composed of approximately 9% mineral spirits and 0.515% cypermethrin and a 35% combination of propane, butane and isobutane. The sole purpose of the propellant is to spread out and distribute the active ingredient, cypermethrin. In essence, the inert ingredients of the product account for 99.485% and the active ingredient accounts for 0.515%.

In addition to the above, major manufacturers, in every one of their applications to the EPA for approval purposes, indicated that their products incorporated only the active ingredients and not the propellants as the ingredients to be evaluated. In fact, the applications submitted stated that the inert ingredients, mainly propellants, amounted to 99.3% or more.

From a formulation standpoint, everything that is in the container for any specific use is the total amount of each ingredient as retailed to the public. For example, in all the applications of a TRF by more than one major manufacturer, propane, butane and isobutane should have been incorporated as part of the application process to the EPA and not designated as an inert ingredient. If it was an inert ingredient the thousands of explosions, fires and resulting injuries over the last 30+ years would never have occurred.

The Propellant

Manufacturers, in most instances, fail to place the EPA and the consumer on notice that the highly flammable and explosive gaseous propellants are invisible and much heavier than air. They also travel along the lower part of the floor unless disturbed. This creates a hidden danger to the consumers but is well known to the manufacturers. More importantly, those facts have been concealed from the EPA in the application and approval process.

What makes it more difficult to detect is the fact that the explosive formulation is invisible and not odorized. In addition, the vapor density of the propellant and active ingredients are heavier than air and tend to fall to the lowest part of a room or floor. In many instances, the vapor can actually flow down a stairway and wind up at the lowest part of the floor below which is usually the furnace. Contrary to the manufacturer's inferences, the TRF is not distributed evenly throughout a room from floor to ceiling. In addition, the system does not follow the ideal gas laws.

The flashpoint for the subject propellant, as stated on the Material Safety Data Sheet, is (minus) -140°F. The flashpoint of a volatile liquid is the lowest temperature at which it can vaporize to form an ignitable mixture in air. Thus, the propellant can ignite at temperatures much lower than normal room temperatures.

Propane has a molecular weight of 44.10 grams per mol as opposed to 28.97 for air so it is about 1.5 times heavier than air. Iso-butane and butane have a molecular weight of 58.12 grams per mole so it is about 2 times heavier than air. When released in air, the propane / isobutene propellant has a negative buoyancy and thus falls under the effect of gravity.

In his textbook "Evaluation of the Effects and Consequences of Major Accidents in Industrial Plants" (Publisher: Elsevier Science, 2008, Chapt. 6, pp. 195-248), Joaquim Casal discusses the flow of heavier than air gasses including propane. The difference in the molecular weight and the negative buoyancy and gravity cause the propellant to fall to the ground level and displace the air. The significant difference in the molecular weight does not allow for the homogeneous mixing of the propellant gas with air.

Aerosols released from a pressurized can will first move in the direction of release due to the momentum of the gas. (Momentum is a quantity in mechanical engineering expressing the motion of a body or system, equal to the product of the mass of a body and its velocity. If the direction of release is pointing upward, the gas will first move up until it loses its momentum under the effects of gravity. At that point, "if the material released is heavier than air, it will first descend to the ground and then spread radially" [Casal page 219].

If the heavier-than-air gas is released in an indoor space, the edges of the cloud will eventually reach the perimeter of the space and then conform to the shape of the bottom section of the space. As Casal states in his book, the mathematical models which predict the actual shape of the cloud as a function of time are complex and require a computer to solve. Most of these models do not account for obstructions in the room. Complications are added if there is any air flow in the room (from a fan, air conditioner, or open window) and if there is a temperature variation in the room from floor to ceiling (because hot air rises). What is known however, is that regardless of the exact shape of the edges of the cloud as time progresses, after several seconds of dispersion, the heavier-than-air gas will displace the air at the bottom of the room and fill the space.

The TRF industry relies upon J. D. Commerford's article entitled: "The Measurement of Propellant Obtained from the Discharge of Total Release Aerosols" published in the July, 1992 issue of the trade magazine "Spray Technology & Marketing." This article could not have been peer-reviewed because the scientific method used was incorrect and inaccurate. The author assumes that the TRFs follow the ideal gas law and the product is distributed throughout each room evenly from floor to ceiling. He also assumes that normal ventilation within all of the environments will dissipate the concentration explosive gases to render them harmless.

Dry air is treated as an ideal gas for scientific analysis even though it is made up of nitrogen, oxygen, carbon dioxide and several other trace gasses. Much of this is done to simplify the analysis and there are many chemical processes that rely on extracting the different gasses from air. Thus, in no case can mixtures of dry air and other gasses be considered to be a uniform ideal gas. As Casal states in his book and illustrates using measurement data from several sources, heavier-than-air gasses have a negative buoyancy in air. When released into the air, heavier-than-air gasses tend to fall under the influence of gravity. The uniform dispersion of the heavier-than-air gasses in an enclosed room cannot occur even after the long periods of time reported.

The heavier-than-air gasses will conform to the floor of the space, displace the air and the volume occupied will be equal to the volume of the heavier-than-air gasses. If the room has a large cross sectional area, the heavier-than-air gasses will occupy a height extremely close to the floor of the room. My expectation when I review the test data reported in the article is that the measurement equipment cannot measure the gas concentrations this close to the floor.

While *not completely incorrect*, the article is certainly misleading. For example, the author of the article says that, "the volatile propellants rapidly flash off leaving finely divided particles of concentrate" [page 29] and "the propellant flashes off into the enclosed air space" [page 28]. The use of the non scientific term "flash off" seems to indicate that the propellant somehow disappears from the room. This is misleading in that the propellant disperses but in no way disappears from the room until a window or door is opened or it seeps through some other opening.

The author of the article also states that there are three schools of thought about where the propellant goes, either falling to the floor, layering itself or being distributed uniformly distributed [page 28]. Again, this is very misleading since it implies that there is disagreement about what happens when a gas is released into an enclosed space. The real situation is that any of these occurrences are possible depending on the properties of the gas.

The author of the article also states, "Instantaneously the volume concentration of propellant in the air are not uniform, but in time the propellant is distributed to all parts of the room." [page 29] This is certainly true but he does not state the obvious implication of this, namely ***that the concentration of the propellant at a certain instant of time and at a particular location of the room can easily be between the lower explosive limit (LEL) and the upper explosive limit (UEL). At this point in space and at this instant of time, if the propellant comes in contact with an ignition source, IGNITION WILL OCCUR.***

There is also a problem with Cummerford's measurement technique in the distance from the propellant source. The sample points were positioned several feet from the source of the propellant. You can model what the exact shape of the plume is as the propellant leaves the can but for the most part, it will not be as you expect. The propellant will go upward because of its momentum as it leaves the nozzle of the pressurized can. Since the propellant is heavier than air, at some height gravity will overcome the momentum and the propellant will begin to fall to the floor. To measure the concentration of propellant where it may be between the lower explosive limit (LEL) and the upper explosive limit (UEL), you would want to put the sample locations within this plume. By placing the sample points several feet from the discharge site, the plume has a large cross sectional area and the concentrations of propellant will clearly be low.

There is a review an article entitled, "Propellant Distribution from Total Release Aerosols Discharged into an Enclosed Chamber, by Richard G. LeVasseur and J. D. Commerford which appeared in the April, 1997 issue of the trade magazine, "Spray Technology & Marketing". This is an update of the previous article for a different propellant and using a different Foxboro instrument for measurement. This article has the same problems as the previous one, namely that the measurement starts too late (although earlier than the previous article) and at a distance that will obviously yield low concentrations.

Lastly, Cummerford did not attempt to test the area with an ignition source because it would have been obvious to the undersigned that there would have been an explosion resulting in injuries to him and others. If that occurred, he would not have been able to write the article for the trade magazine.

It should be noted that the management of the defendant manufacturer used that information for their own product and was testified to be their technical director and research and development manager employed by the defendants. As stated previously,, the assumption that their product follows the ideal gas law is totally incorrect. This can also be demonstrated by taking, for example, a GLADES air freshener and expel some it into a room. The observation of the flow of the ingredients and the vapor follows Casals analysis in the referenced publication.

Chemistry Review of the TRF

A chemistry review of the documents of the manufacturer's reports that the highly flammable explosive repellent under pressure is "Combustible." The manufacturers also state that "exposure to temperatures above 130°F may cause bursting." That is a common statement used on all products containing the propellant. The inference is that the product is not extremely flammable and/or explosive. The technical review branch of the EPA in 2006, reported that the label does not need the signal word: combustible.

The EPA relies on the expertise of the manufacturer to present a complete, accurate and true description of the chemistry, formulation and summary of the testing by the manufacturer.

ANSI Z129.1 1982 Standard titled For Hazardous Industrial Chemicals – Precautionary Labeling states that, "Each hazardous chemical must be considered individually in the light of its own characteristics and customary or reasonably foreseeable handling or use. In such cases, suitable statements should be developed to fit the situation." This is the basic standard of care that the major manufacturers failed to follow.

Since there are millions of organic chemicals and formulations, there is no government agency or international agency that is capable of evaluating each and every formulation. They rely completely on the alleged integrity and honesty of the presenter of the application and/or the manufacturer. In every case, the EPA relies on the manufacturer's integrity of their application in the approval process. They do not have the wherewithal or the facilities nor the time to test the formulations of thousands of applications. In fact, there is no government agency or international agency that is capable of evaluating each and every formulation submitted for approval.

A major corporation's research and development laboratory performed a "Container Stability Determination, Real Time and Accelerated Stability" in their EPA application. The tests also used the "Product Properties Guidelines, Group B-Physical/Chemical Properties." Even though the corporation defined it as a pressurized liquid, not once did they identify the propellant as an extremely flammable/explosive combination of gases when used in a foreseeable protocol. The manufacturers did not properly test the subject product and presented false information in their report.

In one of the cases the undersigned was involved in, the manufacturer reported that the "flammability" is "0" under the section "Hazardous Materials Identification System." The propellant was defined as a "hydrocarbon propellant" with CAS Numbers 75-28-5 and 106-97-8. These gases are a combination of a highly explosive and extremely flammable gas. In section 4 titled "Fire and Explosive Hazards Data," the defendant manufacturer designates that the- "Flammable Limits" is" N/A". One can conclude that it is highly probable that these statements in the MSDS is both fraudulent and false.

More than one manufacturer states that an odor neutralizer is used. This ingredient must also be listed as a component as part of the application submitted to the EPA. The undersigned could not find one document listing the odor neutralizer ingredient.

In addition, manufacturers' MSDS for their TRF products which are described as a total release indoor aerosol insecticide, is formulated with more than 90% of isobutane, butane and propane and is housed within a pressurized container. Under Section III, the MSDS indicates that the vapor density is greater than 1, when in actuality, it varies from 2 to 4. The defendant manufacturer also stateed under "Flammable Limits," that the product is not flammable. That communication to anyone using their products is totally false, misleading and fraudulent.

The EPA's response to all of the misleading information resulted in a suggested hazard statement by them on the label reading, "Contents under pressure. Do not store near heat or open flame. Do not puncture or incinerate container. Exposure to temperatures above 130ºF may cause bursting." There was no mention of the propellants being extremely flammable and explosive because the manufacturer never placed the EPA on notice of that fact and the consumers, with that limited information, released the insecticide throughout the areas of their residence.

Other documents submitted to the EPA by the manufacturer stated that the active ingredients were tetramethrin and cypermethrin. The total of the inert ingredients as listed were 99.30%. The inert ingredients were not defined or identified. However, the manufacturer knew that when their product was released into a room, the "inert ingredients" combined with the active ingredients and distributed the active ingredients within a specified area. In order to obtain the true toxicity of the product, the pressurized liquid concentrate should have been identified. As such, and as elaborated in the "Executive Summary" submitted to the EPA, the information would be questionable. There were many other areas in the EPA application where the flammable characteristics of the repellent, once released, were not properly reported and listed by the manufacturer as not applicable (N/A).

The EPA

The EPA has continually ignored their own experts,, as well as fire officials who are concerned about the potential dangers of explosions and/or fires caused by routine usage or storage of consumer aerosol products. Instead, the EPA responds to pressure from the manufacturers of these products who are resistant to changing warning labels. Extrapolating from data the New York City Fire Department collected concerning the increase in fires and explosions caused by aerosols, an internal EPA memo estimated an excess of 500 fires and/or explosions annually in this country."

In the 1970's, scientists discovered that the chlorofluorocarbons (CFCs) used as propellants in aerosol cans (among other things) were a leading factor in the depletion of the ozone layer. As a result, hydrocarbon propellants (i.e. propane, butane, isobutane) and ether propellants replaced CFCs in aerosol products. The down side to this alternative is that hydrocarbons are extremely flammable substances, and it appears that the public has not been sufficiently warned and educated about their safe use. The

Environmental Protection Agency (EPA) has significant evidence that the propellants can pose significant hazards in normal household use,and yet they have not revealed this information to the public. It is up to the EPA to respond to consumer protection needs. Unfortunately it appears that it reacting to pressure from aerosol manufacturers. The EPA should resist this pressure and change warning labels on many aerosol products, such as total release foggers and space-spray pesticides to warn consumers of the potential dangers.

In response to this danger and an increase in the number of accidents caused by aerosol products, in 1987 the EPA proposed new regulations to improve warning labels on them. But their proposed changes were opposed by the aerosol industry. In an EPA memo one official states, "Industry, through the CSMA [Chemical Specialties Manufactures Association], indicated that they would initiate a lawsuit against PR 87-8 [the EPA's recommended labeling requirements] if it were not rescinded." After much debate, the government succumbed to the legal threats and withdrew their proposal.

The business lobby has been able to cut off all warnings that could aid in decreasing the number of fires and explosions caused by aerosol products. An internal EPA memo states:

"Incident data from the CPSC [Consumer Product Safety Commission] indicates that 60% of pesticide aerosol incidents resulted in explosions that caused injuries requiring medical attention. . . many incidents have resulted due to abuse of the aerosol container.However, if the aerosol cans were obviously marked as 'Flammable' or 'Extremely Flammable' some of the abuse might be prevented. It should be noted that the "abuse" described is foreseeable and known to the manufacturers.

Aerosols using the highly flammable/explosive propellant are used daily in bathrooms, kitchens, and garages, In most cases, the user has no knowledge concerning the propellants in the spray can. According to an article published in Aerosol Age magazine, "Well over half of all aerosols are marked 'FLAMMABLE' or 'EXTREMELY FLAMMABLE'. . .". The article goes on to state, "90% [of aerosol products] contain flammable propellants and over two out of three contain flammable or combustible concentrates." The increase in the number of fires and explosions is do to the increased use of the combustible hydrocarbon propellants.

Aerosol containers are regulated by three different federal agencies: the Environmental Protection Agency (EPA) which regulates pesticide products, the Consumer Product Safety Commission (CPSC) which regulates non-pesticide products, and the Department of Transportation (DOT) which manages bulk shipments of aerosol containers. The DOT has become very effective in regulating this industry. Test methods and labeling seem to be a high priority at the DOT. But the EPA and the CPSC have not had the same attitude toward this issue. While the EPA is the major focus of this report, the CPSC virtually has actually ignored the same consumer safety concerns (i.e. proper testing methods and improved labeling). The EPA in the past has made some effort to look into this issue.while the CPSC has done little to solve the problems aerosols have caused. Ironically, the CPSC is the one federal agency that collects all of the reports of fires and explosions resulting from aerosol containers and still does nothing to improve the protection of consumers. One must remember that it is the CPSC, as well as the EPA and the industry itself that are at fault for aerosol products accidents caused by the innocence of an uniformed consumer.

The EPA began looking into the problem created by chlorofluorocarbons (CFCs) in the mid-1980's. In 1987, the EPA released PR Notice 87-8, "SUBJECT: Label Improvement Program - Required Labeling for Products Containing Flammable Propellants".

PR Notice 87-8 proposed to change the labeling of aerosol products by adding the words, "PHYSICAL OR CHEMICAL HAZARDS" to products that contain propane, butane, isobutane, and dimethyl ether as one or more of the propellants.

This statement was followed by an extended warning label that gave information about use, storage, pilot lights, electrical equipment, incineration and high temperatures. Manufacturers were given until July 31, 1988 to change labels on products released for shipment and July 31, 1989 for all products sold to the consumer.

In response to this initiative, industry dug in their heels. In an EPA memo one official states, "Industry, through the CSMA, indicated that they would initiate a lawsuit against PR 87-8 if it were not rescinded." After much debate PR Notice 87-8 was withdrawn on December 10, 1987, less than ten months after it was published. The EPA released PR Notice 87-11 rescinding the former, "based on the lack of sufficient anecdotal accident information to define the scope of the problem with respect to flammability of aerosols containing the aforementioned propellants". The aerosol industry cast enough uncertainty and the issue to terminate the proposed changes. The termination of the proposed labeling was to be followed by an extended period of data collection and research by the EPA and the industry.

Why has the aerosol industry taken such a stance against the EPA and the consumer? A major reason for the aversion to the previously proposed changes revolves around profits. Changing testing methods and labeling will cost the industry due to the redesigning of aerosol products. Changes would alter the current way the production process occurs and the industry would have to brar the cost of the new labels. Prior labeling would have to be replaced in a specified period of time to be in compliance with regulations and to avoid any penalties.

An EPA memo described a conversation an agency official had with an official from S.C. Johnson & Son, the manufacturers of Raid. According to the EPA official, after discussing testing methods, the industry official stated, "his company does not need to worry about the effects of flammability products since its insurance company settles on any damages or injuries." This statement was also verified by an internal EPA Office of Pesticide Programs memo, "Industry does not appear to be concerned as their insurance companies pay all claims."

How many fires and/or explosions are enough to cause the industry to change its approach to the use of aerosol products? So far it appears that the industry relies on the excuse of "consumer misuse" as the primary reason for accidents caused by aerosol products.. Industry should ask itself why this abuse occurs and how the misuse can be prevented. Instead the manufacturers keep trying to find possible ways to get around the system. According to EPA officials, ways of getting around the system are known, "With current approved testing methodology, a 'nonflammable' rating of the hydrocarbons and ethers can be achieved by proper engineering of the nozzle and the delivery system." Montford A. Johnsen, Technical Editor for Aerosol Age magazine, agrees with this statement. "The nature of the flame projection test invites the formulator to make modifications of product or package to pass it and thus avoid labeling the product as 'Flammable'." Yet consumer misuse is industry's answer to the problem, even though they have been notified of numerous accidents that have resulted because of an uninformed consumer.

Why would the aerosol manufacturers contimued conduct business in this manner? The propellants in question (propane, butane, isobutane and ethers) are some of the least expensive propellants on the market. Alternative propellants would cost more which could eventually drive up the final cost to the consumer.

By providing a full explaination of the contents and the explosion/fire potential, the consumer might become more sensitive to the product and its inherent dangers.For the last eight years the EPA has tried to re-educate the public about the safety risks that could save lives and property. We believe changes are needed before the accident databases overflow with victims. This process may cost aerosol manufactures a small amount at first but in the long run it will save lives.

However, it should be industry's responsibility to educate the public if they cannot use existing technology to make their products safer. The problem is that technology is available that will make the product safe. Therefore, the exisitng products would be considered defectively designed.

Aerosol Propellant for TRF Insecticide and Health Hazards

In addition to the propellant system for TRF insecticides being extremely flammable and explosive, users are also exposed to other health hazards. Exposure to the propellant, once it is released, can cause drowsiness, have an anesthetic effect, cause eye burns and skin burns (frostbite), central nervous system depression, asphyxiation, headache, nausea, unconsciousness, rapid breathing, visual disturbances, muscular weakness, tremors, narcosis, eye damage, mild intoxication, loss of coordination, airways dysfunction syndrome, difficulty in breathing and death. The target organs are the central nervous system, lungs, eyes and skin.

Since the vapor density of the aerosol system is much heavier than air, the propellant tends to stay as close to the floor as possible. When that occurs, there is a lack of oxygen or air in the area adjacent to the floor and no insect is able to survive in that environment. Therefore, contrary to the claims that the defendants made to the EPA that the propellant is considered an inert ingredient, it is very effective in actually suffocating insects on contact.

The use of the propellant system in killing insects by suffocation was not considered by any of the defendants. This is a serious omission and should have been divulged in their EPA application. Furthermore, all of the testing on rats and other animals should have incorporated the use of the propellant rather than using a non-pressurized formulation. As stated in another section of this report, this is further evidence of why the testing performed by the defendants on animals was defective. Their application to the EPA for approval of the efficacy of the active ingredients.was not complete and, therefore, defective as to the results reported and/or not reported.

Asphyxiants, the TRFs and Risk Analysis

Propane, butane and isobutene, for example, are asphyxiants. There is no question that the lower flammability limits of propane and air was exceeded causing both an oxygen-deficient and explosive atmosphere. Butane is also heavier than air as well as isobutene. Inhalation of this combination of gases may cause rapid respiration, dizziness, fatigue and nausea. The manufacturers could have looked up these symptoms on the Internet. There is no question, in the opinion of this author, that they knew of the chemical, physical and health hazards associated with the propellant they used.

In addition to the above, the undersigned consulted on a litigated case in Florida about fifteen years ago where a professional applicator using a similar product wound up with neurological and respiratory injuries. The records for this case were unavailable to the undersigned.

Prior to creating the TRF insecticides, the manufacturers knew or should have known that the standard of care, and custom and practice in the chemical industry that involves hazardous products is that the manufacturer performs a risk analysis on the foreseeable uses and misuses of their product prior to placing it in the stream of commerce. If there is nothing available from a technical standpoint that would have made that product safer or reduce the risk to the user, then they had a duty to warn and place the user on notice of the inherent hazards and dangers associated with their product and methods of protecting themselves and others from foreseeable and possible harm. Based upon the documents reviewed by the undersigned, there is no evidence that the manufacturers performed a risk analysis and properly warned users against the foreseeable dangers and hazards.

If the manufacturers had performed a risk analysis, they would have known how to eliminate the fires and explosions that have regularly occurred over the last four decades. In almost all cases it was just a matter of time, from a statistical basis that the fires and explosions would occur. This is well exemplified by the thousands of explosions that have occurred over the last several decades up to the present time.

If they had researched methods of reducing the risk of the user being exposed to fires and explosions, they would have known that there were safer alternatives available.

Consumers

Consumers were expected to know that the propellant had gone after a certain period of time. This is not within their ability to do so based upon the physical and chemical nature of the product and the environment that the product was used in. The consumer would have had to use an explosion proof exhaust system placed at the doors with all the windows open in order to remove as much of the explosive propellant as possible. Because there are areas in their homes in which the propellant may be trapped, only a professional would know how to safely remove the explosive vapors. A professional would also be able to measure the concentration of the trapped or isolated propellant in any area of the home. The manufacturers are fully aware of thar fact but not the consumer.

The CPSC

The CPSC and litigated cases document the fact that these formulations explode and injure individuals. Through the use of the Freedom of Information Act, one can obtain the incidents without identifying the individuals or parties involved. However, the reports on the Internet estimate that there are over 500 fires and explosions in the United States. Further, the pesticides in bug bombs are also potentially toxic to humans and pets. Rid-a-Bug is fully aware of those reports. The EcoSMART website reports that there are at least 500 fires and explosions resulting from the use of bug bombs. That website was also available to every manufacturer and their management.

When an accident occurs involving a fire or an explosion, often the manufacturers of that product do not make a concerted effort to determine what occurred until litigation is initiated. It is at that time that they blame the plaintiff for misusing their product.. It should also be noted that when there is a fire or explosion where there is minimal damage or injury, those incidents are very seldom, if at all, reported.

In addition, there were no documents available whereby any of the defendants reported accidents and injuries as required to the CPSC. According to their own paperwork, they relied upon self-reporting by those injured. Since the late 1980's that has been unreliable and not a proper measure of the actual accidents occurring with the use of their products.

The defendants in this case have known for decades of reports of fires, destruction of entire apartment buildings (Salt Lake City, September 2009), explosions and burned victims resulting from exposure to propellant systems used in hair sprays, room deodorizers, repellent sprays for mosquitoes insects and other similar products. The only difference is that the other products use very little of the propellant for each application.

In addition to fires and explosions, there are also serious side effects affecting one's health. Consumers have experienced coughing, shortness of breath, airway dysfunction syndrome and upper airway irritation. Insect bombs usually contain pyrethroid and/or pyrethrins as active ingredients in addition to the extremely flammable and explosive aerosol propellants. Not one of the products is natural and in some cases of injury and death, the users followed instructions and vacated their premises for hours. In spite of this, there is one known death from insect bombs involving an infant found dead the day after the insect fogger was used in the infant's home.

The fact that the manufacturer did not list the whole formulation including the propellant system as part of the formulation in their application to the EPA and the public is, in my opinion, misleading and fraudulent. That EPA was never informed of the highly flammable and explosive nature of the propellant system in any of the applications. However, manufacturers, in their applications using the non-flammable propellant, made sure that the agency knew that it was "non-flammable." Their statements and applications speak for themselves.

It is a known fact that the courts do not recognize the self-reporting of injuries because most companies do not make an effort to know what occurred with each and every TRF that was sold. In addition, there is no evidence that manufacturers have reported any incident involving their products to the CPSC. In many instances, the lawyer representing the injured party is the one who makes inquiry.

Warnings and Instructions

The warnings and instructions placed on the subject products are defective. It is difficult to believe that the defendants claimed federal exemption when they misled the EPA with incorrect and fraudulent data regarding the extreme flammability and explosive characteristics of the Total Release Indoor Aerosol Insecticide.

The product does not include hang tags that incorporate the phrase **"DANGER DO NOT RE-ENTER UNTIL THE ROOM/AREA IS COMPLETELY VENTILATED WITH NEW AIR AND ALL FLAMMABLE/EXPLOSIVE GASES ARE ELIMINATED."** This tag should be placed prominently on the entrance to a room or house restricting entry until the area bombed is completely aired out. The hang tag statements would be in both English and Spanish.

In addition to the above, the warning on the subject product should also place the user and others on notice that the product is too dangerous to be used by anyone other than a professionally trained person. The warnings should include the following:

!DANGER!

THIS TRF FOGGER IS EXTREMELY FLAMMABLE AND EXPLOSIVE

THE PRODUCT MUST BE APPLIED BY PROFESSIONALLY TRAINED AND CERTIFIED TECHNICIANS

DISABLE ALL ELECTRICAL PRODUCTS, HOT-WATER HEATERS, FURNACES, REFRIGERATORS, STOVES, PILOT LIGHTS, SPACE HEATERS, OTHERWISE AN IGNITION SOURCE WILL IGNITE THE EXTREMELY FLAMMABLE AND EXPLOSIVE INVISIBLE VAPORS

IF THE PRODUCT IS APPLIED BY ANYONE BUT A CERTIFIED AND TRAINED TECHNICIAN, THERE IS A HIGH PROBABILITY THAT A FIRE AND/OR EXPLOSION WILL OCCUR RESULTING IN SEVERE BURNS AND POSSIBLE DEATH

AFTER 5 HOURS OPEN ALL WINDOWS AND DOORS AND APPLY EXPLOSION PROOF EXHAUST FANS FOR AT LEAST ONE HOUS TO MAKE SURE THAT THE EXPLOSIVE VAPORS ARE ELIMINATED

In addition to the above, a pictogram of a flame and explosion should also be placed on the label. The requirements for the pictogram can be found at 40 CFR 156.78(d).

Another condition that is not mentioned but is an important factor is the weather. In extremely cold climates, is difficult for the product to be used because one would have to open up the windows and doors of the residence, apartment or house, in order to completely eliminate all of the extremely flammable/explosive propellants. The problem is that the gaseous propellant is invisible and only a trained certified technician would be able to measure and determine whether all of the propellant has dissipated and is eliminated.

The Home Insect Control--The Manufacturer's Hype

Dr. Michael F. Potter of the Department of Entomology, College of Agriculture, University of Kentucky, reported in the paper titled, Limitation of Home Insect Foggers ("BUG BOMBS") that, "while foggers are convenient and require little effort or knowledge to dispense, they seldom are effective against indoor pest problems." He goes on to report that "when applied in this manner, very little insecticide actually penetrates into cracks, voids, and other hidden locations where cockroaches, ants, silver fish, and most other household pests congregate and spend most of their time. The ingredients within "Bug Bombs also tend to be repellent, causing insects to scatter and move deeper into wall voids and other hard-to-reach areas."

It is a known fact among entomologists that when insecticides are needed to eliminate pest infestation results are much better if a spray, bait, dust, etc., is placed directly into the areas where the pests are likely to be hiding. They have found that targeted insecticide applications are much more effective against all varieties of household pests, including fleas.

In addition, when using TRF insecticides, the resulting pesticide residues wind up on exposed surfaces and constitute a health hazard. Occupants with asthma and other respiratory ailments react severely when, for example, pyrethrins and other irritating volatile compounds are used indoors.

Safer Alternatives

The technology for safer alternatives has been known for decades. In fact, room deodorizers and hair spray accidents involving fires and explosions using similar propellant systems as the TRF insecticides have been reduced significantly by changing the propellants to non-flammable gases or with the use of manual sprays.

DuPont's Dymel, as previously is a non-flammable propellant system that the defendants knew about and actually applied for an EPA registration approval of the same formulation as the subject product in this case using the DuPont propellant in the year 2000. In spite of this, they continue to use the highly flammable and explosive propellants which ended up severely burning consumers and destroying property.

It should be noted that several or more of the major manufacturers made application to the EPA using DuPont's Dymel 134A. They identified the actual propellant as part of the total formulation. Dymel, although toxic like the combination of extremely flammable and explosive ingredients used in the subject product, is non-flammable and inert. Major manufacturers all knew that this product was available and could be used safely by the general public.

Spectracide, distributed by Spectrum Group, a Division of United Industries Corp., manufactures a non-flammable insecticide that kills ants, roaches, spiders, fleas and ticks. In the EPA registration # 9688-176-8845, the active ingredient is lambda-Cyhalophrin. This product is disbursed by manual spraying and it more than documents the fact that manufacturers using the propane/butane/isobutane propellant had safer products available that would have eliminated the risk of fires and explosions.

The ORTHO Group manufactures and distributes an insecticide called Home Defense Max. The product is non-flammable and will not explode when exposed to an ignition source. The product kills ants, roaches, spiders, fleas, ticks, earwigs, silver fish, crickets, carpet beetles, firebrats and moths. The EPA registration number is 239-2663. The active ingredient is bifenthrin.

Hot Shot Pest Control Concentrate is also a non-flammable, one-ounce product and can be diluted with water to create up to two gallons. The active ingredient is opernethrin and the EPA registration number is 73049-154-8845.

It might suggested that selling TRF insecticides in smaller sizes and as individual "bombs" would reduce the risk. Is a smaller "bomb" safer than a larger "bomb?" The answer is definitely "No." The consumer is still disbursing the explosive and extremely flammable propellant. The smaller size will not significantly reduce the risk.

Conclusions

FIFRA §3(d)(1)(C) provides for an insecticide to be classified for restricted use. The EPA is concerned about unreasonably adverse effects on the environment.

EPA regulations at 40 CFR 152.170 present a more detailed criteria for making a FIFRA §3(d)(1)(C) determination for an end-use product. When deciding whether to classify a product for restricted use, the Agency considers the toxicity of the product along with other pertinent evidence of hazard and, if any serious hazards could reasonably be mitigated by restricting use. They examine whether labeling could adequately address those hazards. Many manufacturers do not divulge all of the risks and hazards associated with the product which is, in this case, the extreme flammability and explosive nature of the propellants used. The fact that many manufacturers made false, misleading and fraudulent statements in their application to the EPA including all of their testing data, should eliminate any claim they would make as to being federally exempted in their product labeling. It is obvious in all cases involving their EPA applications that they intentionally failed to place the EPA on notice by identifying the "inert ingredients" as being non-flammable as described above.

In addition, safety should never be an option with reference to a product being federally exempted. Manufacturers intentionally compromised the safety and welfare of consumers when they could have easily used the non-flammable propellant systems available for their product. Manufacturers should never place TRF insecticides using highly flammable/explosive propellants in the stream of commerce where laypeople can purchase it. They are not trained in the safe use and application of a highly flammable and explosive product.

The EPA was never given the opportunity to weigh the seriousness or magnitude of the relevant risks in order to consider whether restricting the use would be an appropriate means of reducing those risks. The benefits do not outweigh the risks when there are safer alternatives available to the general public.

Author Information:

C. J. Abraham, PhD, DSc, JD, P.E., FRSC, FTI, DEE (IH), BCEE has spent the last 40 years consulting internationally on a variety of personal injury cases. He has also invented and commercialized a variety of products. Over the last 40 years he has evaluated many household and industrial products with reference to their foreseeable uses and foreseeable misuses. He has also tested a large number of products for their flammable and explosive characteristics and created a number of warnings and instructions as well. He is certified and has diplomat status in a number of areas. His background and experience can be found at www.scientificadvisory.com.

(36d) Tips From 25 Years of Environmental and Product Liability Litigation

Gayle S. Koch, Axlor Consulting LLC, Cambridge, MA

Chemical engineers, by choice or chance, often become involved in litigation proceedings. Experience with various environmental and product liability litigation leads to lessons on how to be aware of litigation issues and better prepare for the experience. This session will cover concepts such as economic benefit, allocation of costs, quantification of costs under uncertainty, cost causation, disclosure, and risk assessment issues. In addition, general tips such as standards, documentation and communications will be discussed.

(63a) Preliminary Technical Feasibility Study Into Effect of Withdrawing Spent Caustic Stream From Existing Activated Sludge Treatment Plant

Janit Pongpuak[1], Wiwut Tanthapanichakoon[2], Vasimon Ruanglek[1], Nakarin Mongkolsiri[1] and Surachate Chalothorn[1], (1)SCG Chemicals, Central R& D, Technology Centre, Bangkok, Thailand, (2)Department of Chemical Engineering, Tokyo Institute of Technology, Japan

Our olefins plant produces olefins (ethylene, propylene, butylenes) as well as aromatics (Benzene, Toluene, and Xylene, or BTX) by thermally cracking naphtha in the presence of dilution steam. The plant employs an activated sludge (A/S) wastewater treatment plant to treat a combination of the following three major wastewater streams: the spent caustic stream (SCS) from the cracked gases scrubbing unit, the blowdown stream (DSG) from the dilution steam generator used to produce steam for the thermal cracker, and the dilution water stream (DWS) consisting of rainwater, equipment cleaning water, fire alarm testing and etc. As part of our waste minimization policy, a new plan is afoot to divert the SCS completely to a caustic soda recovery plant, thereby not only significantly reducing the total influent flow to the A/S plant but also changing the wastewater characteristics. This leads to a necessity to conduct a technical feasibility study into the corresponding effects on the performance of the A/S plant. In practice it is also necessary to know how the SCS flow diversion should be implemented in order to ensure smooth withdrawal of SCS during the transition period.

Simultaneously, the feasibility of total elimination of the DWS and furthermore the possible inclusion of wastewater from a nearby HDPE plant in the near future are to be investigated. To smoothly accomplish the above-mentioned series of planned changes, a series of preliminary experimental studies at the lab-scale level (using batch reactors and continuous reactors) have been carried out to study the A/S process performance under different scenarios. Based on the experimental data, a dynamic math model of the lab-scale A/S process is also developed and substantiated. In the next step the extension of the math model to the existing A/S plant will be carried out whilst taking into the difference in residence time distribution in the lab-scale and full-scale aeration tanks. Then the validated model will be used to predict the A/S plant performance in response to the planned withdrawal of SCS and DWS streams as well as the subsequent inclusion of the HDPE plant stream. What will be reported here is the lab-scale experimental results and the corresponding math model substantiated by these experimental results.

(63b) Cleaner Chemical Industry Through HEN Advanced Retrofit Approach

Mahmoud Bahy Noureldin, Process and Control Systems, Saudi Aramco, Dhahran, 31311, Saudi Arabia, Mohammad Alibrahim, Process and Control systems Department, Saudi Aramco, Dhahran, Saudi Arabia, Mana Al-Owaidah, Process and Control Systems, Saudi Aramco, Dhahran, Saudi Arabia and Ali AlQahtani, Process and Control systems, Saudi Aramco, Dhahran, Saudi Arabia

The economics of industrial production, the limitations of global energy supply, and the realities of environmental conservation are an enduring concern for all industries. Wherever you turn, there's another entreaty to save energy, reduce carbon emissions and protect the environment for posterity.

Industrial plant's heat exchangers network (HEN) shall always be ready to face a high level of operational changes along its lifetime. These changes are short-term, such as process disturbances; uncertainty in feed stock conditions and product demand, and long-term such as the need to process more raw materials that warrants the debottlenecking of the facility including the HEN to increase its capacity. Nowadays, and since late seventies of the last century, another important frequent change is happening due to the continuous escalation in energy prices in a rate that is higher than plant equipment cost, also warrants continual modification of the facility's HEN to enhance the plant energy efficiency along its lifetime (sometimes reaching up to 30 years). In such cases, the HEN retrofit objective/task is to produce a practically implementable cost effective HEN design modification that satisfies the new process objective and its new operating constraints. There are many possible modifications for an existing HEN to retrofit the original design to the new objective. It can include a combination of all possible process operating and design conditions modifications, existing HEN topological/structural modifications, and existing HEN unit design modifications and parametric modifications (such as heat transfer enhancement to improve U) as well. In huge industrial facilities such problem can be solved using pinch technology or MINLP , where we have an objective to minimize total cost of the existing HEN modifications and the rest of the plant's process units design and operation modifications minus the realized energy cost saving. There are some known drawbacks in the existing methods. In one hand the most widely used to date pinch method and its modifications is non-systematic and involves heuristics that produce solutions whose quality depends entirely on the designer experience in applying the heuristics. For instance, determining the location of the pinch point for an assumed best globalDT_min, (which globalDT_min should be used) is a little arbitrary. As the value of assumed best globalDT_min changes, the location of the pinch point changes and therefore the assessment of which unit to transfer heat across the pinch also changes. The same also applies upon process changes that result in same output. The retrofit is a one package solution that has no consideration for yearly step-by-step modifications and practicality for future retrofit projects. The mathematical programming-based methods on the other hand have the drawback of providing little or almost no scope for user interaction. Nowadays, it is possible to solve larger problems than before but still not on the scale needed in many industrial mega sites. The problem of the HEN retrofit is NP-hard, and including all possible process design modifications, different types and configurations of heat exchanger units and so on in the superstructure is impractical and will make the problem harder, intractable and impossible to solve. In addition, the MINLP model cost objective function needs comprehensive data base for the model solution to be realistic with information/data that is normally not available during conceptual phase of the retrofit project, especially in de-centralized engineering environment where projects cost control depends entirely on rigorous cost calculation that always comes after the retrofit project basic engineering and is carried out at different departments in most of the companies. Besides neither practical nor logical to assume with high fidelity all possible retrofit solutions that can be implemented for a given plot plan of a facility and rigorously develop cost correlation to all of it before running the mathematical program especially when there is always more than one option to carry out a designated modification such as the increase of the network surface area. Neither one of these methods (pinch technology, network pinch and mathematical programming) can rigorously assess the existing network; rigorously target for the solution before the go ahead of the retrofit project for practically attainable waste heat recovery and/or minimum number of units that need to be added to reach desired energy consumption targets or part of it, and to systematically renders solutions that can be implemented on phases with a guarantee that the solution implemented today in phase one will not become an obstacle, or need to be demolished, for a tomorrow retrofit solution to be implemented at later time in the future on phases two, three and so on along the plant's lifetime. The paper exhibits new elements, in the representation and assessment of the existing HEN studied for retrofit and in the systematic finding of the retrofit solutions that is evolutionary in implementation and revolutionary in consideration. It presents the concept of plant life-time retrofit-ability that addresses the identification of any possible contradiction between now and future retrofits to render opportunities of continuing retrofit projects to achieve ultimate target. The method is illustrated with two industrial case studies for NGL and oil refining plants.

(63c) Energetic Self Sufficiency: Theory and Application

Vasilios Manousiouthakis, Chemical & Biomolecular Engineering Department, UCLA, Los Angeles, CA and Jorge Pena Lopez, Chemical and Biomolecular Engineering Department, University of California, Los Angeles, Los Angeles, CA

Abstract

The concept of energetic self-sufficiency, for an open thermodynamic system, is first presented and its application to hydrogen production is discussed.

Let Ω be a steady-state open thermodynamic system with inlets in S_I; outlets in S_O; participating chemical species in S_C; participating elements in S_E; heat transfer rates Q_j with j in S_Q entering ($Q_j>0$) or exiting ($Q_j<0$) the system at temperatures T_j respectively, and shaft work rates W_j, j in S_W entering (consumed by) the system ($W_j>0$) or exiting (produced by) the system ($W_j<0$). This system has to satisfy the overall mass, element, energy, and entropy conservation laws. These latter two laws are often referred to as the 1st and 2nd laws of thermodynamics, which at steady-state require that the energy entering the system must equal the energy exiting the system, and that the total entropy change of both the surroundings and the system is greater than or equal to zero. Then

Definition. The system Ω is called energetically self-sufficient iff no heat is transferred from the surroundings to the system, heat is possibly transferred from the system to the surroundings ($Q_j<0$, j in S_Q) at the uniform surroundings temperature T_0, no shaft work is utilized (consumed) by the system, and possibly shaft work is generated (produced) by the system ($W_j<0$, j in S_W).

The considered application of this concept involves the thermodynamic assessment of hydrogen production from natural gas so that no carbon dioxide is emitted from the system. The concept of thermodynamic self-sufficiency is used to establish rigorous limits on the amount of hydrogen that can be produced.

(63d) Impact of Flaring Emissions on Regional Air Quality Associated with An Ethylene Plant Start-up

Jian Zhang, Qiang Xu and Thomas C. Ho, Dan F. Smith Department of Chemical Engineering, Lamar University, Beaumont, TX

Flaring is a common industrial practice designed to eliminate the accumulation of un-reacted feed gases or off-specification product gases through stack burning to ease the strain on a process unit especially during plant start-ups, plant shut-downs and operational up-sets. Although the practice is necessary and crucial to chemical plant safety, excessive flaring represents the loss of precious feed materials and also results in the generation of extra amounts of undesired emissions including VOCs and NO_X, both of which are ozone precursors.

In this study, the effects of flaring emissions associated with an ethylene plant start-up with various start-up scenarios on the increment of regional ozone concentrations were investigated through model simulations. In the practice, the dynamic process simulations associated with an ethylene plant start-up were performed first. The obtained flaring emissions were then incorporated into a CAMx air quality simulation model to simulate the effect of these extra VOCs and NO_Xemissions on the regional ozone concentrations. From the the CAMx simulations, the effects of start-up starting time, plant location, and episode day on ozone increments associated with various start-up scenarios were investigated and scientific insights were also provided. Finally, the environmentally-preferred start-up schedule is identified as the one that avoids having the peak flaring emissions occur near the time of the highest solar intensity to minimize the impact of the start-up process on the regional ozone concentrations.

(63e) Phytoremediation of A Wetland SOIL, by TWO METHODS In Bogota Colombia

Ingrid Rivera[1,2], Sergio Pedraza[1] and Gabriel Camargo[3], (1)Environmental Engineering, Universidad Libre, Bogota, Colombia, (2)Universidad Libre, Bogota, Colombia, (3)Ingenieria Mecanica, Universidad Libre, Bogota, Colombia

In this project, the efficiency of two phytoremediation methods for soil (phytoextraction and phytostabilization) was investigated to remove heavy metals such as Pb, Cu and Ni. The place for this research was a wetland call "La conejera".

In this wetland the pollution levels has been growing due the urbanization impact in Colombia. The first step in the project was physicochemical characterization, followed by the selection of the phytoremediation process. The methods chosen to remove Pb, Cu and Ni, were phytoextraction and phytostabilization. The ability and effectiveness to remove the heavy metals was evaluated.

For the phytoextraction process, the mustard (sinapis Alba) and barley (*Hordeum vulgare)*were the plants selected. The mustard was used due its adaptability in wet soils, faster germination and growth. The barley was used due resistance to extreme weather conditions. The results obtained shown that all metals were remove by the barley and only the Copper when the mustard was used.

In the phytostabilization method, the plant used was Ryegrass perenne (*lolium perenne L.*). The results obtained shown that all metals was retired for this plant.

2012 AIChE Spring Meeting &
Global Congress on Process on Process Safety
April 1-5, 2012 - Houston, TX

Search of Suitable Halon 1301-Like Fire Suppressants Based on Flame Inhibition Mechanism Understanding

C. Osorio[1], E. Petersen[2] and M.S. Mannan[1]

Mary Kay O'Connor Process Safety Center, Artie McFerrin Department of Chemical Engineering, MS-3122, Texas A&M University, College Station, Texas USA
[2]Department of Mechanical Engineering, Texas A&M University, College Station, Texas USA
Corresponding author: epetersen@tamu.edu

ABSTRACT

Halon 1301 (CF_3Br) had been widely used as a fire extinguisher for many years due to its unique properties including high extinguishing efficiency at low concentrations, high dispersion capabilities, and storage stability, among others. Unfortunately, CF_3Br and other halogenated fire suppressants have been restricted by the Montreal Protocol due to their high ozone depleting potential. This restriction has led to the necessity to find cleaner alternatives with similar, Halon 1301-like capabilities. Despite all the scientific efforts made during the last few decades, no single suitable Halon 1301 substitute has been found. A more likely approach to overcome this issue is to study a variety of novel choices that can be used on specific situations. Identification of these alternatives requires a systematic methodology based on the understanding of the flame inhibition mechanisms that could address this search in promising directions. To this end, the present work presents a methodology to investigate fundamental phenomena involved in the combustion inhibition mechanism of CF_3Br by using detailed chemical kinetics which have been experimentally validated. Specialized combustion equipment, including a high-pressure shock tube and a flame speed bomb, are used to obtain important parameters including ignition delay time, OH* time histories, and flame speeds which are also employed as metrics for model validation. In particular, laminar flame speed is considerably important since it represents a fundamental property of the reacting mixture and provides a tool for deeper flame inhibition analysis to evaluate thermal and chemical inhibition contributions and saturation effects. Sensitivity analysis is performed to generate valuable information regarding key elementary reactions that limit and control flame inhibition behavior. Halon 1301 may be unique in its ability to modify both the chemical kinetics and thermal properties of a reacting system. With a fundamental understanding of the chemical and thermal advantages of CF_3Br as a fire suppressant, the same fundamental tools can be employed to test for similar attributes in candidate compounds. This approach provides a guide for identification of novel, chemically active inhibitors which behave similar to CF_3Br but can be significantly less harmful to the environment.

1. INTRODUCTION

Halon 1301 (CF_3Br) had been widely used for many years due to its unique properties including high extinguishing efficiency at low concentrations, high dispersion capabilities, and storage stability, among others. Unfortunately, CF_3Br and other halogenated fire suppressants have been restricted by the Montreal Protocol due to their high ozone depleting potential [1-3]. It led to the necessity to find cleaner alternatives with similar, Halon 1301-like capabilities. Despite all the scientific efforts made during the last decades, researchers have not been able to find a single, suitable Halon 1301 substitute. Identification of likely alternatives requires a systematic methodology based on flame inhibition mechanistic understanding that can address this search in promising directions.

To this end, the present work presents a methodology to investigate the effect of fire suppressants. This paper shows part of an overall research effort that started with the examination of CF_3Br. Therefore, potential candidates can be tested and compared with Halon 1301, which is used as reference.

2. FLAME INHIBITION ANALYSIS: METHODOLOGY

The proposed methodology, shown in Figure 1, is applied to investigate fundamental phenomena involved in the combustion inhibition mechanism of different flame inhibitors, starting with CF_3Br. Then, similar properties can be sought in other fire suppressant candidates/blends using similar analytical procedure. To accomplish this, model formulations and experimental designs are integrated in order to examine the chemical kinetics and thermal effects of fire suppressants on different hydrocarbons, especially methane, ethane and propane. Comparison between numerical and experimental combustion parameters, such as ignition delay times, OH* time histories, and flame speed are used as metrics for mechanism validation. In particular, laminar flame speed, considered a fundamental property of the reacting mixture in premixed flame, will be used for flame inhibitor characterization, but also to examine the thermal and kinetic inhibition mechanism of a specific substance or blend. In addition, identification of the key elementary reactions and pathways that most affect inhibition mechanism are determined through sensitivity analysis.

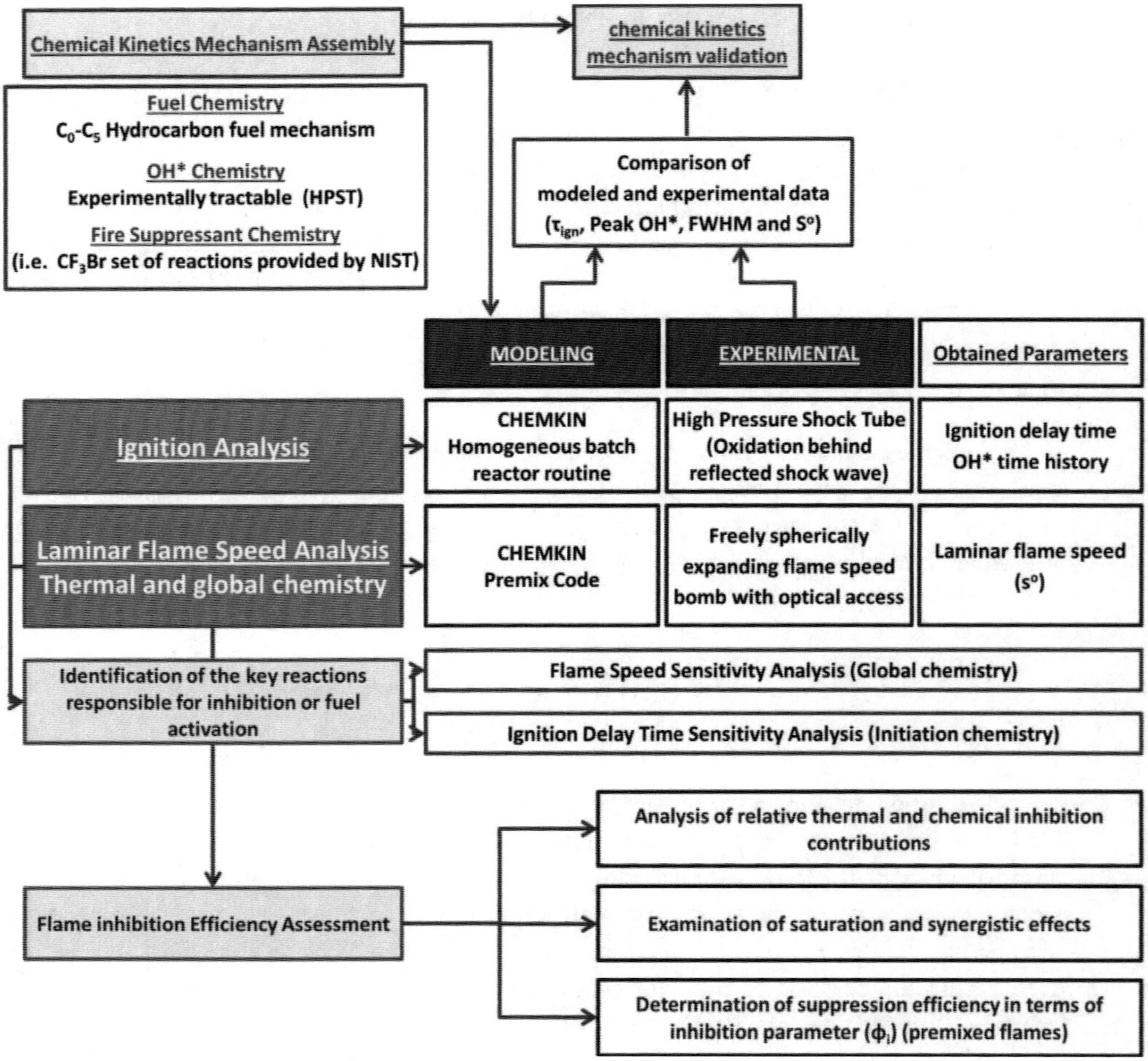

Figure 1. Methodology for flame inhibition analysis

2.1 Modeling

Modeling predictions are conducted using CHEMKIN. A detailed chemical kinetics mechanism was assembled from three existing sets of chemical reactions: (1) a C_0-C_5 hydrocarbon mechanism from NUI-Galway [4], (2), the Hall and Petersen OH* sub mechanism [5], and (3) Halon 1301 chemistry which have been provided by National Institute of Standards and Technology (NIST) [6]. Use of this model provides novel information regarding fire suppressant effects on a wide range of fuels (C_0-H_5), as well as OH* chemistry that can be directly compared with experimental ignition results. The complete mechanism, used for Halon 1301 study, includes 333 species involved in 1980 reactions.

2.2 Experimental

Examination of fuel ignition using a High Pressure Shock Tube: Effect of CF_3Br on the fuel ignition is analyzed using a High Pressure Shock Tube (HPST). The facility is provided with driven and driver sections that have ID's of 15.24 cm and 7.62 cm, respectively, both with a 1.27-cm wall thickness, constructed entirely of stainless steel 304 (Fig. 2). The two sections are separated by a diaphragm that ruptures depending on the pressure ratio between the two sections. The reactant mixture is contained at the driven section (low pressure). It will ignite by the action of a shock wave produced after the diaphragm ruptures. In these experiments, fuel oxidation is reached behind reflected shock waves where zero dimensional conditions are achieved. Reaction progress is monitored using sidewall emission measurements of OH* chemiluminescence at 310 n. Then, ignition delay time (Ţign) can be determined by analyzing the Pressure/OH* emission profiles obtained in each experiment (Figure 3). A more detailed description of this facility and technique is provided by Aul et al. [7].

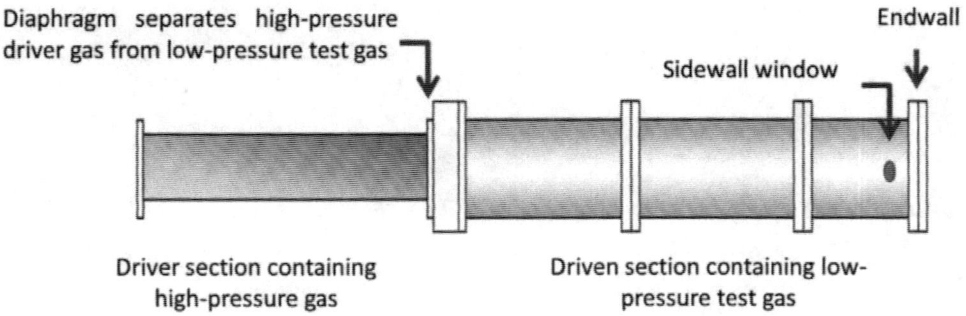

Figure 2. Primary sections and diaphragm station of a simple shock tube. [7].

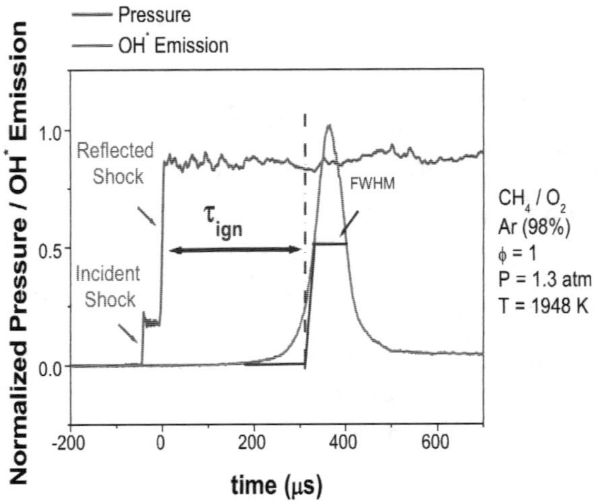

Figure 3. Typical pressure and OH* measurements from the shock-tube experiment

Laminar Freely Propagating Flame Speed: Evaluation of the laminar flame speed can be made by using different techniques, however, most of them involved flame curvature, non-uniform upstream flow, and heat transfer and other effects that makes difficult their modeling [8]. The method used in this study examines freely propagating flames, minimizing complexity of the system (e.g. minimum heat losses). This facility, described in more detailed by de Vries et al. [9], consists of a aluminum alloy constant-volume cylindrical vessel, with an inner diameter of 30.5 cm and length of 35.6 cm. Direct observation of the flame allows tracking the changes of flame radii as a function of time (Figure 4). For this radii tracking, a high-speed camera (up to 20,000 frames per second) is used as element of the Z-type schlieren optical diagnostic technique.

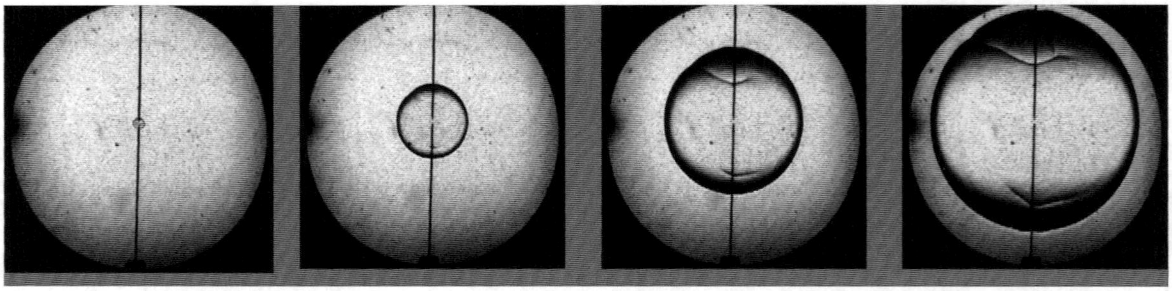

Figure 4. Example of experimental images of an expanding flame.

3. RESULTS

Ignition results: Different mixtures of fuel (methane, ethane and propane), oxygen, and Halon 1301, highly diluted in argon (98% Ar by volume) were evaluated. Experiments were conducted at different temperatures (1250 - 2250 K) and equivalence ratios at atmospheric pressure. Figure 5 shows the effect of CF_3Br on ignition delay time for stoichiometric mixtures. Two different aspects are examined: 1) the effect of CF_3Br on methane, ethane, and propane combustion, and 2) the agreement between experiment and model. Ignition Delay Time results show that CF_3Br speeds up ignition in all methane mixtures tested. On the other hand, CF_3Br acts by retarding the ignition in those mixtures containing ethane and propane. In most of these cases, the model predicts the experimental behavior; but in general, it underestimates the resulting effect, indicating that some improvements can still be made.

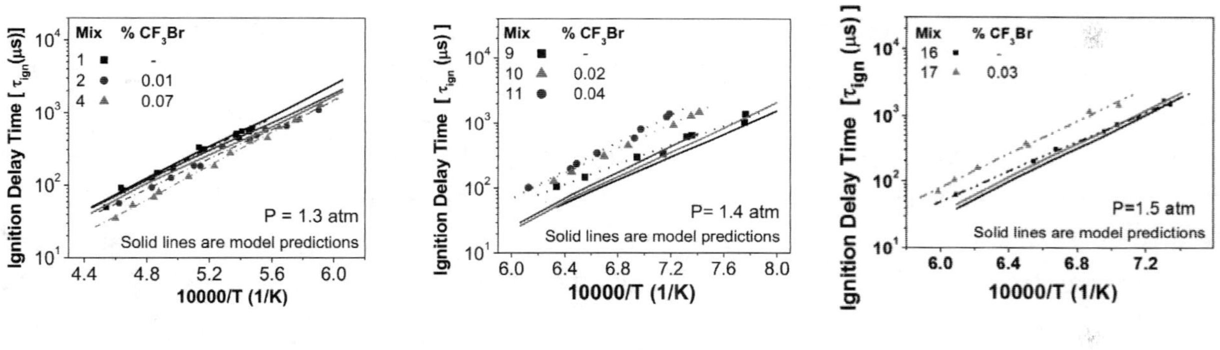

Figure 5. Ignition Delay Time results comparing data to kinetics model for (a) Methane ϕ=1.0, (b) Ethane ϕ=1.0, and (c) Propane ϕ=1.0.

Laminar flame speed results: The effect of CF_3Br on methane, ethane and propane flames was assessed by adding two different concentrations of 0.5% and 1% of Halon 1301 to different fuel/air mixtures, at different stoichiometric ratios from 0.7 to 1.3, at ambient temperature. Figure 6 illustrates the results obtained from experimental and modeling calculations. In all the cases, CF_3Br shows to present an inhibitor effect by decreasing the laminar flame speed. Modeled data foretells same experimental trend, however, improvement in the kinetic mechanism can be made for better predictions.

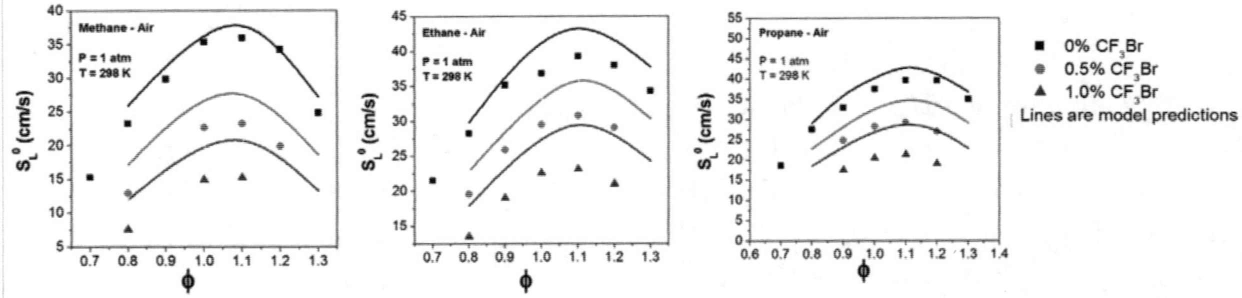

(a) Methane – Air (b) Ethane – Air (c) Propane - Air

Figure 6. Experimental and modeled Flame Speed of different fuel-air mixtures at different CF_3Br concentrations.

<u>Model improvement and identification of key elementary reactions</u>: To improve the model, it was necessary to identify the key elementary reactions that most affect the laminar flame speed in a determined system. Among the reactions that are activated by CF_3Br (reactions involving Br or F), the reaction (R1): Br + HCO => HBr + CO presented the highest negative sensitivity coefficient, for all the systems analyzed. Considering this result, the R1 pre-exponential factor (A) was modified by trial and error. Results illustrated in Figure 7, shows an improvement of methane flame speed predictions by changing the rate of R1 by five Similar results were found for ethane and propane. These results are not aimed to be used as the final and more precise model predictions, but rather show the methodology that can be used for chemical kinetic mechanism improvement.

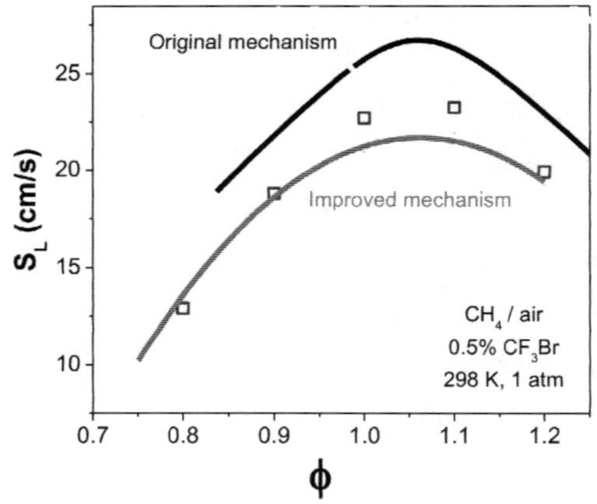

Figure 7. Comparison between experimental (blue points) with the original (black line) and improved modeled (red line) flame speed data for methane mixtures (0.5% CF_3Br, 1 atm, 298K).

Flame inhibition efficiency assessment: Analysis of the relative physical (thermal) and chemical (kinetic) inhibition contribution of CF_3Br was conducted. First, it was necessary to determine the laminar flame speed reduction due to the global effect (physical + chemical) in terms of (S/So) global. For this calculation, the flame speeds with Halon (S) and without Halon (So) were calculated using the mechanism containing F and Br chemistry. Then, an inhibition parameter can be obtained considering the following expression [6], where X_{O2} and X_{CF3Br} are the concentrations of oxygen and Halon 1301, respectively.

$$\Phi = \frac{X_{O2}}{X_{CF3Br}} ln\left(\frac{S_L}{S_o}\right)^{-1} \quad \text{Equation 1}$$

Table 1. Relative chemical and physical inhibition contributions of CF_3Br on methane, ethane and propane stoichiometric flames.

Inhibition Parameter	**Methane**	**Ethane**	**Propane**
Φ_{global}	17.7	9.6	10.8
$\Phi_{physical}$	2.1	2.0	1.6
$\Phi_{chemical}$	15.6	7.6	9.2

Then, the reduction of flame speed by physical means was calculated by isolating the F and Br chemistry from the chemical kinetics mechanism, and calculating the flame speed at this condition. The chemical contribution was obtained by subtracting the physical from the global effect. Results (Table 1) show that CF_3Br behaves as a fire suppressant mainly by its chemical mechanisms. The results, in agreement with previous authors [10-13], demonstrates the capability of CF_3Br to reduce H-radical concentrations in the flame reaction zone, eventually decreasing the rate of the chain branching reaction.

4. SUMMARY AND CONCLUSIONS

Despite the effort conducted toward finding suitable fire suppressants, there is still a need of novel candidates with properties similar to CF_3Br. This search should be based on flame inhibition understanding. Based on this desire, the present work aims to improve the understanding of the Halon 1301 mechanism as a fire inhibitor by employing an analytical methodology capable of examining both thermal and chemical effects on hydrocarbon flames. Similar analysis can be used for evaluating suitable substitutes using Halon 1301 as a reference.

The proposed analytical methodology is based on detailed chemical kinetics mechanisms which can be validated experimentally. The fuel kinetics mechanism employed in this work (C_0-C_5) allows extending the examination of a fire suppressant's effect to more-practical fuels. Specialized combustion equipment, including a high-pressure shock tube and a flame speed bomb were used to obtain important parameters including ignition delay time, OH* time histories, and laminar flame speed. These parameters serve as metrics for model validation and will be tools for deeper flame inhibition mechanism analysis. Sensitivity analysis was performed to obtain valuable information regarding key elementary reactions and pathways. This approach helps to understand the limiting and controlling flame inhibition behavior, and can also be used as a tool for model improvement as well.

5. REFERENCES

[1] J.S. Daniel, S. Solomon, R.W. Portmann, R.R. Garcia, J. Geophys. Res., 104 (1999) 23871-23880.

[2] UNEP, in: U.N.E.P. (UNEP). (Ed.), Montreal, Canada, 1987.

[3] in: U.S. Code (Ed.) VI, 1990.

[4] N. Donato, C. Aul, E. Petersen, C. Zinner, H. Curran, G. Bourque, Journal of Engineering for Gas Turbines and Power, 132 (2010) 051502.

[5] D. Healy, N.S. Donato, C.J. Aul, E.L. Petersen, C.M. Zinner, G. Bourque, H.J. Curran, Combustion and Flame, 157 (2010) 1526-1539.

[6] T. Noto, V. Babushok, A. Hamins, W. Tsang, Combustion and Flame, 112 (1998) 14.

[7] C.J. Aul, DeVries, J. and Petersen, E. L in: Eastern States Fall Technical Meeting of the Combustion Institute, Charlottesville, VA, 2007.

[8] J. De Vries, in: Mechanical Engineering, Texas A&M University, College Station, 2009, pp. 186.

[9] J. De Vries, W. Lowry, Z. Serinyel, H.J. Curran, E.L. Petersen, Fuel, 90 (2011) 8.

[10] C.K. Westbrook, Combustion Science and Technology, 34 (1983) 201-225.

[11] T. Noto, V. Babushok, J. Burgess, A. Hamins, W. Tsang, A.W. Miziolek, Proceedings of the Combustion Institute, 26 (1996) 1377-1383.

[12] V. Babushok, D. Burgess, W. Tsang, in: Halon Options Tenchical Working Conference, 1994, pp. p. 205.

[13] C.R. Casias, J. Thomas McKinnon, Proceedings of the Combustion Institute, 27 (1998) 2731-2739.

OPTICAL FIBER REFLECTANCE PROBE FOR DETECTION OF SUPERCRITICAL TRANSITION

Yujian Sun, Boung Wook Lee, Milorad P. Dudukovic
Chemical Reaction Engineering Laboratory (CREL),
Department of Energy, Environmental and Chemical Engineering,
Washington University in St. Louis, St. Louis, Missouri 63130

Abstract

An optical fiber reflectance probe is developed for the fast, in-situ detection of supercritical transition in multiphase and multicomponent systems. Experiments with two binary mixtures methanol-cyclohexane and CO_2-methanol are conducted to demonstrate the capability of the reflectance probe to capture the supercritical transition in both liquid-liquid mixture and gas expanded liquids. The current probe can withstand pressures up to 140 bars and temperatures of 120 °C, which suggests its applicability in many practical processes.

Keywords

Supercritical transition, reflectance probe, methanol-cyclohexane, CO_2 expanded liquids

1. Introduction

Supercritical fluids, at temperature and pressure above their critical points at which the phase boundaries cease to exist, are suitable substitutes for organic solvents in many processes. Specifically, carbon dioxide expanded liquids (CXLs) are promising as alternative media for performing separations, extractions, reactions, and other applications [1-7]. Although the critical phase transition of a mixture from subcritical to supercritical state can be observed by the naked eyes in vessels with fully transparent walls, that transition is hard to detect in most practical reactor vessels with steel walls unless expensive view cells are provided. High temperatures and high pressures (HTHP) are typically employed in mixed autoclaves for processes where the above transition is expected to occur at some operating conditions. Thus, it is desirable to develop an approach for in-situ detection of the critical phase transition.

Making use of the fact that a transition from subcritical to supercritical state coincides with the change of the fluid from heterogeneous opalescent to homogeneous transparent state, we are able to design and manufacture an optical fiber probe to detect the reflected light which travels a certain known distance in the mixture and therefore carries information related to the transition [8]. We illustrate here the changes in the light signal captured during in-situ detection in the methanol-cyclohexane (liquid-liquid) mixture as the system temperature is raised from below to above critical temperature at fixed pressure. We also illustrate the signal changes with pressure in the CO_2-methanol (gas-liquid) system at different temperatures. These examples illustrate the applicability of the optical fiber probe in the detection of critical phase transition for multiple types of mixtures.

2. Experimental

2.1. Probe Setup

The optical probe system used in this work consists of three optical fibers (I, II and III), as shown in Figure 1. Fiber III (ø1035 μm) is coupled closely with fibers I and II (ø630 μm), which are connected to the laser source (Thorlabs SIFC635PM) and the light detector (Thorlabs PDA36A), respectively. The other end of fiber III, with a small piece of mirror attached about 1~2mm away, is immersed in the mixture being studied. In this way, light coming from the laser source can be transmitted to the mixture through fiber I and III, and then reflected back by the mirror to fiber III and sent back to the light detector through fiber II. The light signal can be converted to an analog signal by the detector, which can then be transferred to the data acquisition system (PowerDAQ PD-NBC-16) and recorded by the computer.

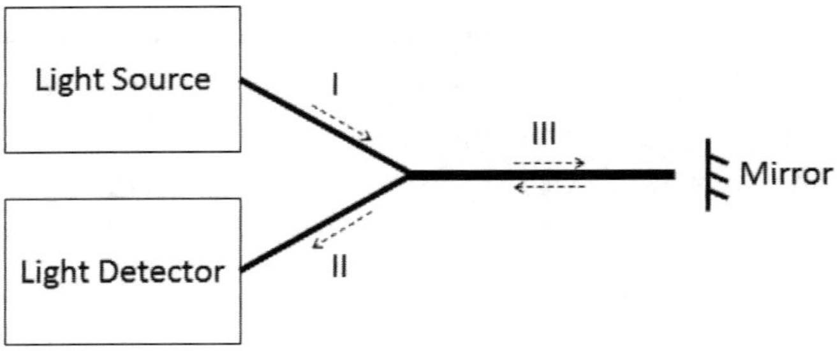

Figure 1. Schematic for the optical fiber probe system

2.2. Rationale

The light signal acquired by the detector consists of two parts, one reflected by the mirror and the other scattered by the mixture. The mirror is attached sufficiently close to the probe end that the reflected part dominates in the total signal. The "noisy" part of the signal due to contribution of scattered light is negligible compared to the intensity of the reflected light from the mirror. That is why we selected the name of reflectance probe.

When a mixture is at subcritical state, there exist distinct interface boundaries between the two immiscible distinctly different phases. Under sufficient stirring, the interface boundaries are broken down into small pieces and distributed randomly throughout the heterogeneous mixture, resulting in strong fluctuation of light signals. During supercritical transition, when interphase boundaries cease to exist, the mixture turns to homogeneous transparent state, thus the fluctuation of the light signal is greatly reduced at the same stirring rate. This disappearance of the fluctuations distinguishes the supercritical state from the subcritical state, and helps in determining the critical conditions.

2.3. Procedure

For methanol-cyclohexane system, as it is stirred and gradually heated up, the temperature change is recorded by video at the same time with data acquisition of the light signal. When the temperature exceeds the critical temperature for the specific composition, there is a signal change which coincides with the transition. From the data obtained, we can find the exact time of this signal change, and since we have a record of the temperature with time we obtain the critical temperature.

For CO_2-methanol system, the experimental setup is shown in Figure 2. CO_2 is pumped into the 3-in. inner diameter (ID), 1-L autoclave. Pressure is controlled via a Tescom 4000 back-pressure regulator and Validyne pressure transducers. Temperature control is achieved using Omegalux heating tape wrapped around the autoclave, an Omega thermocouple inserted into the vessel's thermowell, and an Omega CN132 temperature controller [9]. The end of the optical reflectance probe is immerged in the liquid (methanol) to detect the supercritical transition. As CO_2 is gradually pumped in, the pressure increases and exceeds the supercritical pressure at specific temperature, which results in a signal change that can be captured and analyzed.

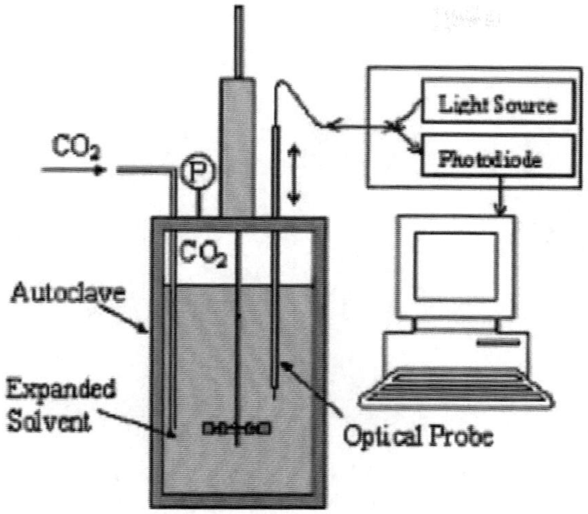

Figure 2. Experimental setup for the autoclave

3. Results and Discussion

3.1. Methanol-Cyclohexane (L-L)

As the liquid mixture of methanol and cyclohexane is heated up to reach its critical temperature, the stirred mixture turns to homogeneous supercritical state, and thus a light signal change can be clearly captured. Figure 3 shows in-situ data collected with 50% mole fraction of methanol. In the same way, critical temperatures of methanol-cyclohexane mixtures are determined at different compositions, and all agree well with previously obtained data [10], as shown in Figure 4.

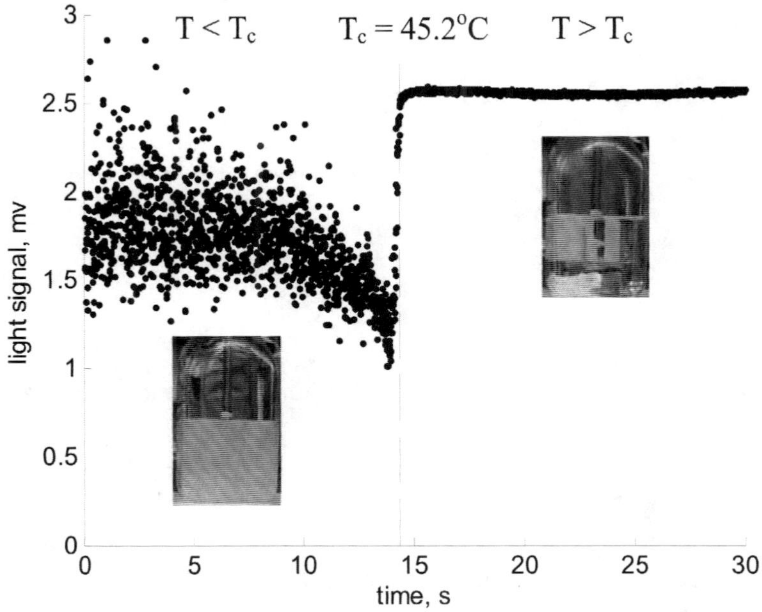

Figure 3. In-situ light signal for methanol-cyclohexane mixture during supercritical transition ($X_{methanol}$ = 50%; Sampling Frequency = 100 Hz)

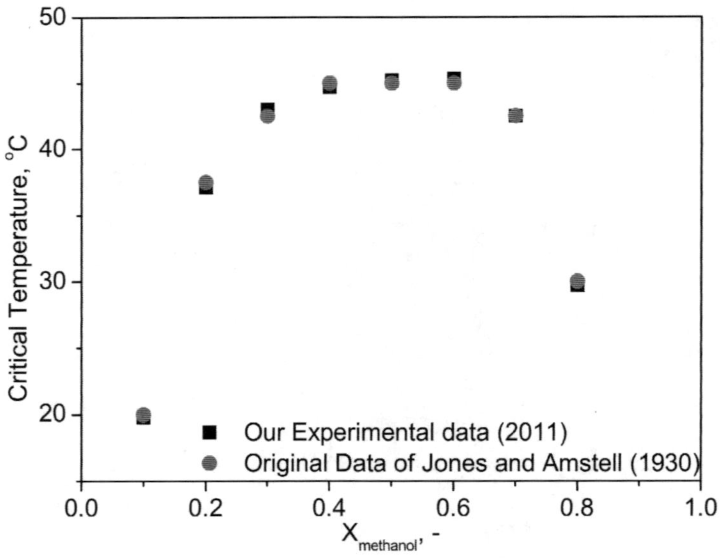

Figure 4. Critical temperatures for methanol-cyclohexane at different compositions

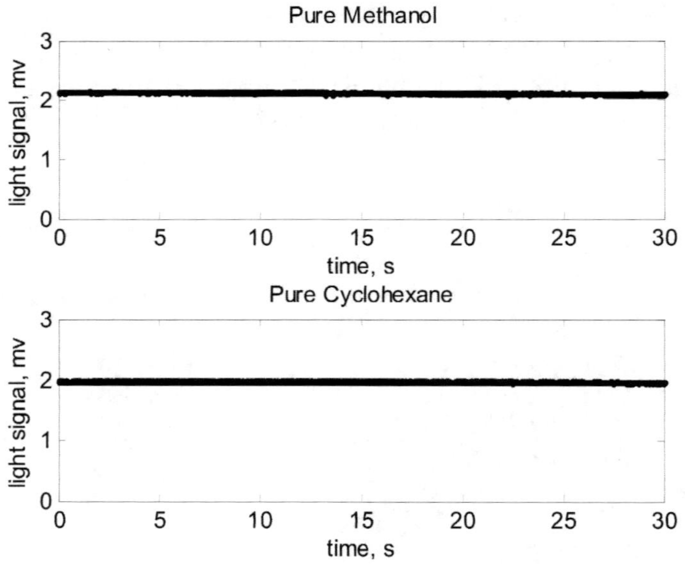

Figure 5. Light signals for pure methanol and pure cyclohexane during heating up

As seen above, at transition point we always find an abrupt rise in the signal. The reason for this change is as follows. In the subcritical methanol-cyclohexane mixture, the heterogeneity results in opalescence as shown in the left photo in Figure 3. This opalescence increases the scattering in the liquid, but it greatly reduces the reflectance since it blocks the path of reflected light and scatters it also. During supercritical transition, the mixture turns from heterogeneous opalescent to homogeneous transparent state, thus the reflected part of the light is enhanced while the scattered part is reduced. Since reflectance dominates in the signal, the entire signal increases sharply at transition.

There is another interesting phenomenon worth noting. The signal intensity continuously and gradually goes down until it reaches a minimum before rising up sharply during transition. Since the same phenomenon regarding detected light intensity cannot be observed for pure methanol and pure cyclohexane, as shown in Figure 5, this has to be related to interactions between the two phases. It seems to indicate that interphase boundaries increase in number during heating up which is responsible for blocking more reflected light and resulting in decreased total signal intensity. Whether that implies smaller diameters of dispersed phase but large numbers requires more study.

3.2. CO_2-Methanol (G-L)

Critical pressures for CO_2-methanol at different temperatures are determined similarly by the same reflectance probe. Figure 6 shows a set of data collected at 25.2 °C. We can still see the disappearance of light intensity fluctuations during supercritical transition. However, the abrupt rise in the signal at transition, which was evident in Figure 3 for the methanol-cyclohexane mixture, is clearly missing in Figure 6 for the carbon dioxide-methanol mixture. The reason might be that this initially heterogeneous gas liquid mixture consisting of small bubbles dispersed in the liquid has different optical properties opalescence character than the methanol-cyclohexane mixture. Our probe can withstand high pressures as shown in Figure 7, which suggests its applicability in many processes.

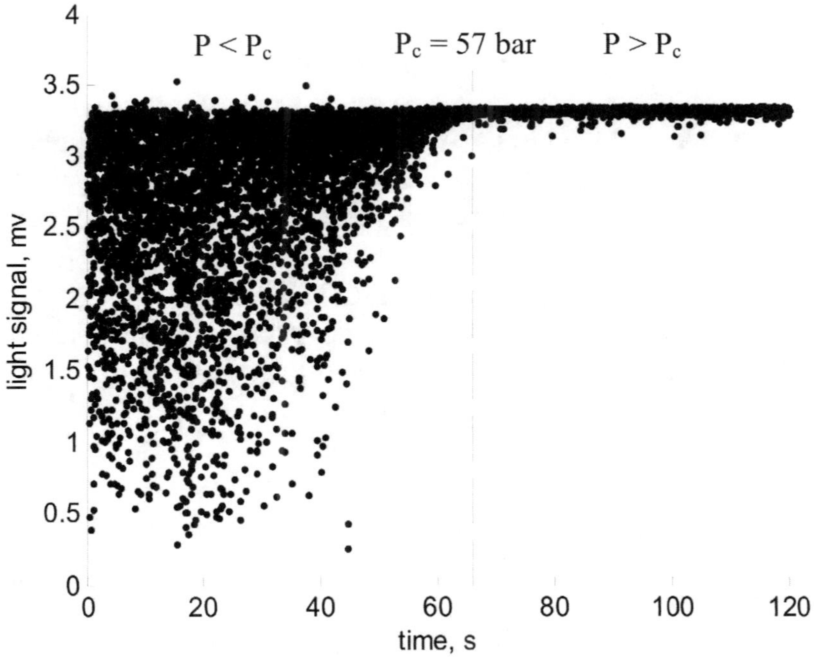

Figure 6. In-situ light signal for CO_2-methanol mixture during supercritical transition ($T = 25.2$ °C; Sampling Frequency = 100 Hz)

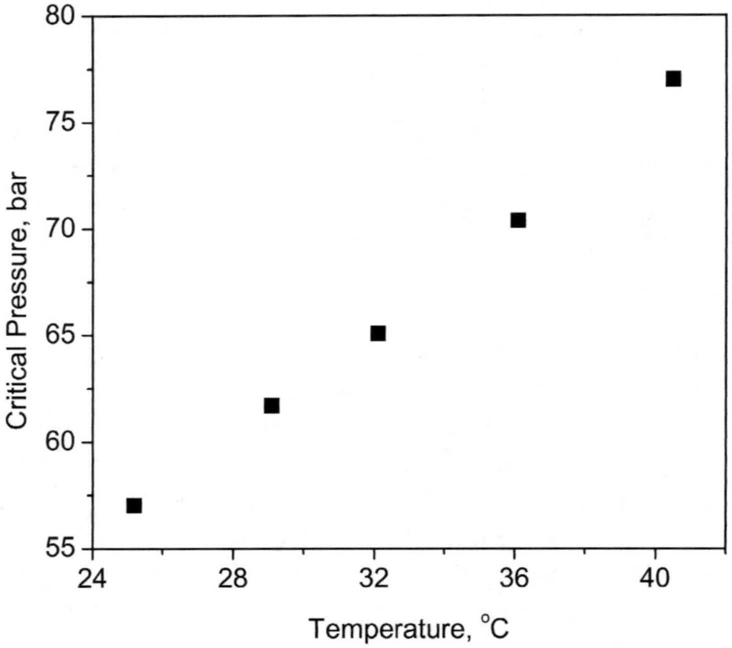

Figure 7. Critical pressures for CO_2-methanol at different temperatures

4. Conclusion

A reflectance probe is developed in this work to detect the reflected light signal change during supercritical transitions of both methanol-cyclohexane and CO_2-methanol system. This signal change can be coupled to the temperature or pressure recorder to determine the critical condition at interest.

ACKNOWLEDGMENTS

This work was supported in part by the grant from the National Science Foundation (CBET 0933780).

References

1. Adams D. (2000), “Clean and green…but are they mean?” Nature, 407, pp. 938-940.
2. Subramaniam B. and McHugh M. (1986), "Reactions in Supercritical Fluids-A Review", Ind. Eng. Chem. Process Res. Dev., 25, pp. 1-12.
3. Savage P. E., Gopalan S., Mizan T. I., Martino C. J. and Brock E. E. (1995), "Reactions at supercritical conditions: Applications and fundamentals", AIChE J., 41, pp. 1723-1778.
4. Jessop P. G. and Subramanian B. (2007), “Gas-Expanded Liquids”, Chemical Reviews, 107, pp. 2666-2694.
5. McHugh M. and Krukonis V. J. (1994), “Supercritical Fluid Extraction”, Butterworth-Heinemann, Boston.
6. Jessop P. G., Ikariya T. and Noyori R. (1999), "Homogeneous Catalysis in Supercritical Fluids", Chem. Rev., 99, pp. 475-494.

7. Wei M., Musie G. T., Busch D. H. and Subramaniam S. (2002), "CO_2-expanded solvents: unique and versatile media for performing homogeneous catalytic reactions", J. Am. Chem. Soc., 124, pp. 2513-2517.
8. Mueller S. G. (2009), "Optical Measurements in Gas-Liquid Stirred Tanks", Doctorate Dissertation, Washington University in St. Louis.
9. Mueller S. G., Werber J. R., Al-Dahhan M. H. and Dudukovic M. P. (2007), "Using a Fiber-Optic Probe for the Measurement of Volumetric Expansion of Liquids", Ind. Eng. Chem. Res., 46, pp. 4330-4334.
10. Jones D. C. and Amstell S. (1930), "The Critical Solution Temperature of the System Methyl Alcohol-Cyclohexane as a Means of detecting and estimating water in Methyl Alcohol", J. Chem. Soc., pp. 1316-1323.

() Musings on Management

Frank van Lier, The Lubrizol Corporation, Wickliffe, OH

It really is all about the people!

Hiring good people.

Retaining them.

Adjusting for different personal styles, cultures, encouraging diversity, embracing different perspectives.

Motivating people to work together in successful global teams.

Achieving top notch results that are better than any individual can achieve alone.

Maintaining the focus.

Getting the results.

Management is all about the excitement managers gets from having their people and teams grow and succeed.

(106a) Synthesis of PVA Based Membranes Modified by Benzimidazole Group for Direct Methanol Fuel Cells

Pratima Gajbhiye, Chemical Engineering, Indian Institute of Technology, Kanpur, State of Uttar Pradesh, India, Jayant K. Singh, Chemical Engineering, Indian Institute of Technology, Kanpur, Ka, India and Anil Kumar, Department of Chemical Engineering, Indian Institute of Technology Kanpur, Kanpur, India

Abstract

In recent years, there has been a great effort to look for alternate membrane material and Polybenzimidazole (PBI) membranes have been synthesized where the conducting moiety of this polymer is benzimidazolium groups. The monomers used to synthesize these polymers are bis conjugated amines and diacids are expensive. In this work, we have synthesized grafted polymers of PVA as membrane materials where the grafts are either Poly Acrylic Acid, Poly maleic anhydride (PMA) or Poly Methacrylic acid (PMAA) and these materials were denoted by PVA-g-PAA, PVA-g-PMA, and PVA-g-PMAA. We then synthesized C1, C3 & C4 compounds (formula given at the bottom of Table S1) having bis conjugated amines whereas compound C2 having only one conjugated amine group. One of the steps in synthesizing C1 compound is nitration of phenyl ring, where there is a possibility of formation of two isomers out of which only one of them has strictly bis conjugated amines and can be polymerized with Isophthalic acid giving the formation of PBI. The other byproduct formed is an impurity and is minimized by carrying out the nitration step at low temperature but this cannot be completely removed from the product which is a mixture. Table S1 presents the summary of our experimental results on characterization of membrane, In the C1 to C4 compounds and in some cases (membranes 9, 10, 11), Sulfocuccinic acid were added to the casting polymer solution as additional crosslinking agent in order to achieve higher conductivity. We added the polymer, thus formed, to the casting solution, we first compared the results of the Fuel cell formed by adding C1 (C3 and C4) compound alone and heating at 180°C for some time. Table S1 has two sets of results, one on characterization (columns 1, 2, 3,4,5,6, and 7) of membranes while the other on Fuel cell (columns 8 to 11). For the latter columns 10 and 11 gave results on characterization on Fuel cell while columns 8 and 9 gave results when the Fuel cell was run on control mode for higher efficiency.

Experiments on membrane pore sizes are given in column 1 which lies between 8.88 to 9.52nm and they all are ultra filtration type. The values of the pore size are mostly governed by the nature of the graft chains, e.g. membranes 1,3,4,7 and 12 consist of PAA grafts and were found to have similar pore sizes. Similarly membranes 2,5,8,9 were of the similar pore size because the PVA have the same grafts of PMA. The next important thing is to determine the wall charges of these membranes and they were estimated by experimentally determining the rejection of Chromic (VI) acid solution through these membranes. Subsequently, a space charge model was proposed for the flow of the solution through their pores and a semi analytical solution was derived for the overall problem of separation of chromic acid solution. We observed that the high surface charge was because of the benzimidazolium groups present on the surface and they ranged from minimum of 465 to maximum of 525mV. The thermal degradation (TGA) experiments were carried out which showed that, through the modification of the polymer by benzimidazole groups, lead to stable membrane and the degradation temperature varies from 150°C for 10% weight loss and for approximately 80% weight reduction, a minimum of 300°C to 700°C.

The twelve membranes reported in Table S1 were first characterized completely in terms of pore size, wall charge, conductivity, IEC and thermal degradation. In fuel cell applications, higher temperatures give large current density and it is observed that some of the membranes (rows 1, 2, 5, 6, 7, 8 and 12) could easily be operated upto 150°C without the membrane degrading. However in our case, we insisted that the membrane of direct methanol fuel cell were operated at 90°C for which most of the fuel cell results for Nafion 117 are available. The catalysts used on the anode and cathode side were Pt-Ru and Pt loaded on carbon respectively, which were synthesized by following procedure given in literature. We find that in each case, the power density and the current density were higher and for some of our results (membranes 1,6,12) were atleast two to three times higher than that of Nafion 117 as seen in column 10 and 11.

It was recognized that the membrane material has hydrophilic PVA matrix and hydrophobic benzimidazolium groups at the surface formed by reaction of the C1 to C4 compounds with grafts (see Fig.S-1). When the reaction occurs (confirmed by IR and NMR), the hydrophobic benzimidazolium are formed which tends to separate as confirmed by the SEM results. In view of this we assumed that the pores of the membrane would have a thin hydrophobic layer with positive charge walls. This is confirmed by the contact angle measurements where the minimum contact angle is 51° and varied upto 89°. Even though most of the polymer matrix is PVA with its grafts, when the ions move through the pores (giving rise to current in Fuel cell), the pore walls get highly charged (as measured by column 2 of Table S1). Some of the membranes have Sulfo succinic acid SSA as part of the matrix (e.g. the membrane 9 to 11) which would tend to give sulfonate groups inside the PVA matrix as well as at the phase boundaries, this way giving rise to bipolar nature of the membrane (benzimidazolium group being strictly cationic in nature). In addition to this, some of the compounds such as C1 and C2 compounds have bonded sulfonate groups which would give the positive and negative charge separated by the molecular distance, once again producing a bipolar nature to the membrane. In remainder of the cases the charges would be induced on the phase boundaries as seen in Fig.S-1. The conductivity observed in our membranes is very high and could be attributed to the special bipolar structure of our polymers.

Table S1. Fuel cell and Membrane Characterization

Compound	Membrane	1	2	3	4	5	6	7	8	9	10	11

			MWCO (pore size, nm)	Wall Charge	Conductivity (S/cm)	IEC	Thermal Degradation temp (°C) 10% - 80%		Water Uptake	Contact Angle	Average Voltage Gain (V)	Average time cycle (min)	Power Density (mW/cm)	Current Density (mA/cm)
C1	1	PVA-g-PAA-C1	9.1	488	0.04	2.7	199.732	268.70	7.5	72	0.523	55.75	50.47	74.88
	2	PVA-g-PMA-C1	9.35	465	1.23×10^{-2}	1.8	142.659	447.6	10.691	74.917	0.568	42.896	0.305	3.81
	3	PVA-g-PAA-modified	9.52	488	1.51×10^{-2}	2.12	115.123	706.53 (28.93%)	13.665	75.581	0.778	69.85	37.37	64.44
C2	4	PVA-g-PAA-C2	9.52	525	6.59×10^{-2}	2.38	119.558	691.93	13.4	51.2	0.704	39.83	23.0020	50.55
	5	PVA-g-PMA-C2	9.36	519	1.04×10^{-2}	2.17	181.59	562.348	14.9	50.24	0.77	161.93	34.72	62.11
	6	PVA-g-PMAA-C2	8.99	482	1.16×10^{-2}	2.18	201.055	472.568	13.8	51.8	0.6066	147.81	76.72	92.33
C3	7	PVA-g-PAA-C3	9.0	512	2.51×10^{-2}	2.23	204.908	453.03	8.624	70.368	0.6883	82.55	36.397	75
	8	PVA-g-PMA-C3	9.5	510	3.51×10^{-2}	2.27	166.819	505.293	13.513	62.944	0.726	37.557	36.11	64.381
	9	PVA-g-PAA-SSA-C3	9.36	497	1.71×10^{-2}	2.51	154.914	813.769 (24.32%)	10.5	54.8	0.8096	99.423	23.45	57.778
	10	PVA-g-PMA-SSA-C3	9.83	480	2.28×10^{-2}	2.46	115.12	428.713	8.5	76.2	0.663	224.22	28.561	58.358
	11	PVA-g-PMAA-SSA-C3	8.88	525	3.66×10^{-2}	1.85	139.772	681.715 (30.78%)	7.2	89.2	0.77	18.36	20.832	48.11
C4	12	PVA-g-PAA-C4	9.5	495	2.56×10^{-2}	1.62	189.87	451.86	8.927	59.132	0.704	74.19	41.1	71
	13	Nafion 117			0.1								21.01	48.33

Compound C1, [Bis-3-Amino-4-{3-(triethylammoniumsulfonato) phenylamino}-phenyl sulfone] hydrochloride

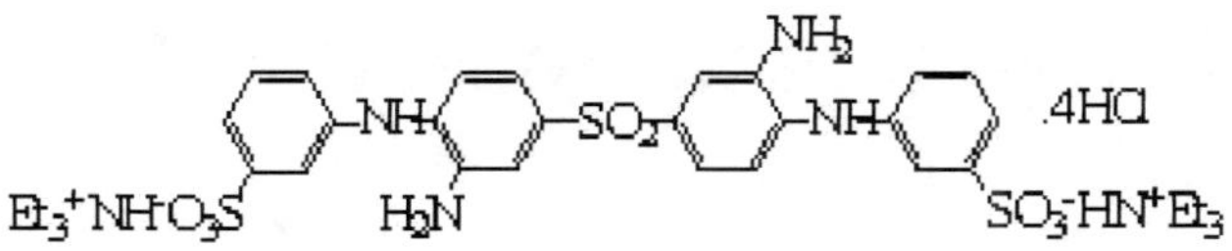

Compound C2, 3-Amino -4-[3-(triethyl ammonium sulfonato) phenyl amino]

phenylene hydrochloride

THE GRAPHIC ON THIS PAGE WAS NOT AVAILABLE ON THE E-MEDIA VERSION OF THIS PAPER.

Compound C3, Bis -4 [1, 2 diamino phenyl]-phenyl sulfone) hydrochloride

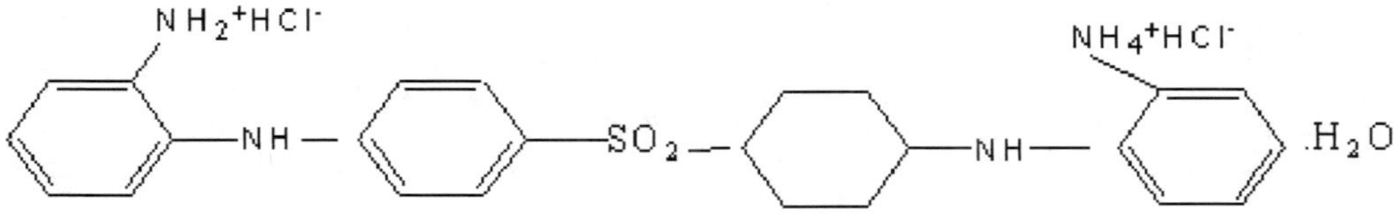

Compound C4, 1, 4 [1, 2 diamino phenyl] Phenylene hydrochloride

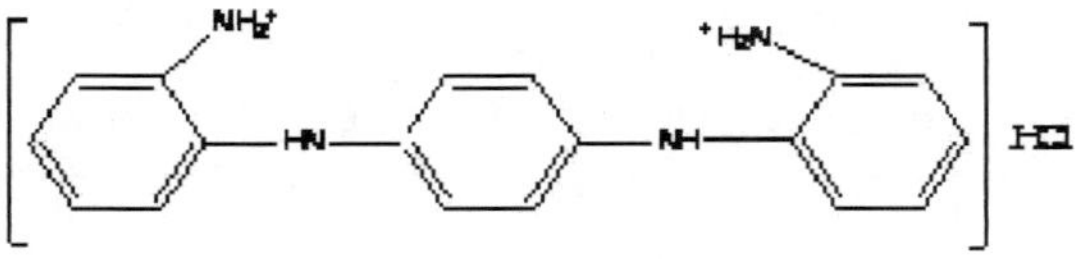

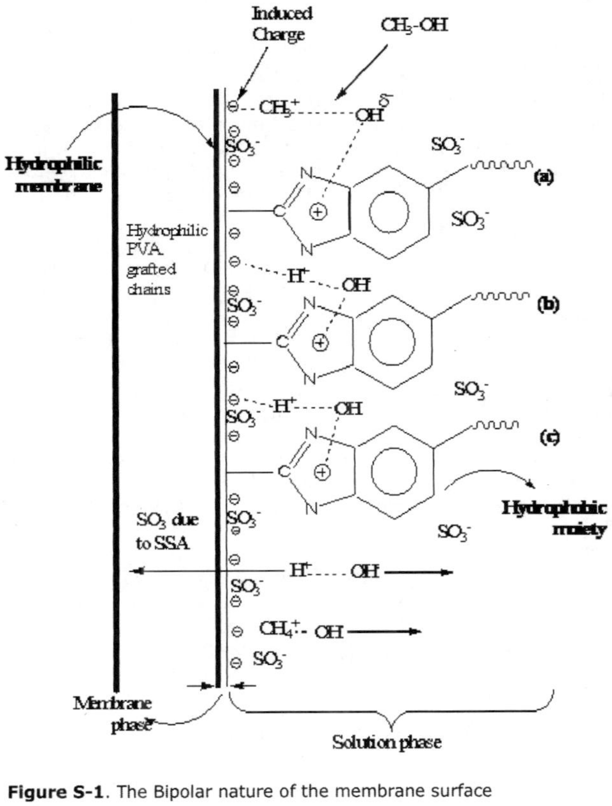

Figure S-1. The Bipolar nature of the membrane surface

(106b) Operation and Controls of Operating Direct Methanol Fuel Cells with PVA Based Membranes Modified by Benzimidazolium Groups

Pratima Gajbhiye, Chemical Engineering, Indian Institute of Technology, Kanpur, State of Uttar Pradesh, India, Pranjal Joshi, Chemical Engg, IIT Kanpur, Kanpur, India, Jayant K. Singh, Chemical Engineering, Indian Institute of Technology, Kanpur, Ka, India and A. Kumar, Department of Chemical Engineering, Indian Institute of Technology Kanpur, Kanpur, India

Operation and Control of Direct Methanol Fuel cell with PVA based membranes modified with benzimidazolium groups

Pratima Gajbhiye, Pranjal Joshi, Jayant K.Singh, Anil Kumar

Department of Chemical Engineering, Indian Institute of Technology, Kanpur 208016, India

Abstract

In recent years, in an effort to find alternate membrane material (other than Nafion 117), Polybenzimidazole (PBI) membranes have been synthesized where the conducting moiety of this polymer is benzimidazolium groups. The monomers used to synthesize these polymers are bis conjugated amines and diacids are expensive. In this work, we have synthesized grafted polymers of PVA as membrane materials where the grafts are either Poly Acrylic Acid, Poly maleic anhydride (PMA) or Poly Methacrylic acid (PMAA) and these materials were denoted by PVA-g-PAA, PVA-g-PMA, and PVA-g-PMAA. We then synthesized C1, C3 & C4 compounds (formula given at the bottom of Table S1) having bis conjugated amines whereas compound C2 having only one conjugated amine group. One of the steps in synthesizing C1 compound is nitration of phenyl ring, where there is a possibility of formation of two isomers out of which only one of them has strictly bis conjugated amines and can be polymerized with Isophthalic acid giving the formation of PBI. The other byproduct formed is an impurity and is minimized by carrying out the nitration step at low temperature but this cannot be completely removed from the product which is a mixture. Table S1 presents the summary of our experimental results on characterization of membrane, In the C1 to C4 compounds and in some cases (membranes 9, 10, 11), Sulfocuccinic acid were added to the casting polymer solution as additional crosslinking agent in order to achieve higher conductivity. We added the polymer, thus formed, to the casting solution, we first compared the results of the Fuel cell formed by adding C1 (C3 and C4) compound alone and heating at 180°C for some time. Table S1 has two sets of results, one on characterization (columns 1, 2, 3,4,5,6, and 7) of membranes while the other on Fuel cell (columns 8 to 11). For the latter columns 10 and 11 gave results on characterization on Fuel cell while columns 8 and 9 gave results when the Fuel cell was run on control mode for higher efficiency.

Experiments on membrane pore sizes are given in column 1 which lies between 8.88 to 9.52nm and they all are ultra filtration type. The values of the pore size are mostly governed by the nature of the graft chains, e.g. membranes 1,3,4,7 and 12 consist of PAA grafts and were found to have similar pore sizes. Similarly membranes 2,5,8,9 were of the similar pore size because the PVA have the same grafts of PMA. The next important thing is to determine the wall charges of these membranes and they were estimated by experimentally determining the rejection of Chromic (VI) acid solution through these membranes. Subsequently, a space charge model was proposed for the flow of the solution through their pores and a semi analytical solution was derived for the overall problem of separation of chromic acid solution. We observed that the high surface charge was because of the benzimidazolium groups present on the surface and they ranged from minimum of 465 to maximum of 525mV. The thermal degradation (TGA) experiments were carried out which showed that, through the modification of the polymer by benzimidazole groups, lead to stable membrane and the degradation temperature varies from 150°C for 10% weight loss and for approximately 80% weight reduction, a minimum of 300°C to 700°C.

The twelve membranes reported in Table S1 were first characterized completely in terms of pore size, wall charge, conductivity, IEC and thermal degradation. In fuel cell applications, higher temperatures give large current density and it is observed that some of the membranes (rows 1, 2, 5, 6, 7, 8 and 12) could easily be operated upto 150°C without the membrane degrading. However in our case, we insisted that the membrane of direct methanol fuel cell were operated at 90°C for which most of the fuel cell results for Nafion 117 are available. The catalysts used on the anode and cathode side were Pt-Ru and Pt loaded on carbon respectively, which were synthesized by following procedure given in literature. We find that in each case, the power density and the current density were higher and for some of our results (membranes 1,6,12) were atleast two to three times higher than that of Nafion 117 as seen in column 10 and 11. In normal Fuel cell operations, there is decay in the cathode catalyst which is further increased by the methanol crossover. The voltage of the Fuel cell increases, the average volatage gained is reported in column 8 and it is found that the voltage gain depends on the nature of the membrane which is not the same for all of them. In column 9, we have measured the average time cycle and we found that some of the membranes (membranes 6, 7, 9, 10) gave extremely good results.

It was recognized that the membrane material has hydrophilic PVA matrix and hydrophobic benzimidazolium groups at the surface formed by reaction of the C1 to C4 compounds with grafts (see Fig.S-1). When the reaction occurs (confirmed by IR and NMR), the hydrophobic benzimidazolium are formed which tends to separate as confirmed by the SEM results. In view of this we assumed that the pores of the membrane would have a thin hydrophobic layer with positive charge walls. This is confirmed by the contact angle measurements where the minimum contact angle is 51° and varied upto 89°. Even though most of the polymer matrix is PVA with its grafts, when the ions move through the pores (giving rise to current in Fuel cell), the pore walls get highly charged (as measured by column 2 of Table S1). Some of the membranes have Sulfo succinic acid SSA as part of the matrix (e.g. the membrane 9 to 11) which would tend to give sulfonate groups inside the PVA matrix as well as at the phase boundaries, this way giving rise to bipolar nature of the membrane (benzimidazolium group being strictly cationic in nature). In addition to this, some of the compounds such as C1 and C2 compounds have bonded sulfonate groups which would give the positive and negative charge separated by the molecular distance, once again producing a bipolar nature to the membrane. In remainder of the cases the charges would be induced on the phase boundaries as seen in Fig.S-1. The conductivity observed in our membranes is very high and could be attributed to the special bipolar structure of our polymers.

Table S1. Fuel cell and Membrane Characterization

Compound	**Membrane**		**1**	**2**	**3**	**4**	**5**		**6**	**7**	**8**	**9**	**10**	**11**
			MWCO (pore size, nm)	**Wall Charge**	**Conductivity (S/cm)**	**IEC**	**Thermal Degradation temp (°C) 10% - 80%**		**Water Uptake**	**Contact Angle**	**Average Voltage Gain (V)**	**Average time cycle (min)**	**Power Density (mW/ cm)**	**Current Density (mA/ cm)**
C1	1	PVA-g-PAA-C1	9.1	488	0.04	2.7	199.732	268.70	7.5	72	0.523	55.75	50.47	74.88
	2	PVA-g-PMA-C1	9.35	465	1.23×10^{-2}	1.8	142.659	447.6	10.691	74.917	0.568	42.896	0.305	3.81
	3	PVA-g-PAA-modified	9.52	488	1.51×10^{-2}	2.12	115.123	706.53 (28.93%)	13.665	75.581	0.778	69.85	37.37	64.44
C2	4	PVA-g-PAA-C2	9.52	525	6.59×10^{-2}	2.38	119.558	691.93	13.4	51.2	0.704	39.83	23.0020	50.55
	5	PVA-g-PMA-C2	9.36	519	1.04×10^{-2}	2.17	181.59	562.348	14.9	50.24	0.77	161.93	34.72	62.11
	6	PVA-g-PMAA-C2	8.99	482	1.16×10^{-2}	2.18	201.055	472.568	13.8	51.8	0.6066	147.81	76.72	92.33
C3	7	PVA-g-PAA-C3	9.0	512	2.51×10^{-2}	2.23	204.908	453.03	8.624	70.368	0.6883	82.55	36.397	75
	8	PVA-g-PMA-C3	9.5	510	3.51×10^{-2}	2.27	166.819	505.293	13.513	62.944	0.726	37.557	36.11	64.381
	9	PVA-g-PAA-SSA-C3	9.36	497	1.71×10^{-2}	2.51	154.914	813.769 (24.32%)	10.5	54.8	0.8096	99.423	23.45	57.778
	10	PVA-g-PMA-SSA-C3	9.83	480	2.28×10^{-2}	2.46	115.12	428.713	8.5	76.2	0.663	224.22	28.561	58.358
	11	PVA-g-PMAA-	8.88	525	3.66×10^{-2}	1.85	139.772	681.715	7.2	89.2	0.77	18.36	20.832	48.11

		SSA-C3						(30.78%)						
C4	12	PVA-g-PAA-C4	9.5	495	2.56×10^{-2}	1.62	189.87	451.86	8.927	59.132	0.704	74.19	41.1	71
	13	Nafion 117			0.1								21.01	48.33

Compound C1, [Bis-3-Amino-4-{3-(triethylammoniumsulfonato) phenylamino}-phenyl sulfone] hydrochloride

Compound C2, 3-Amino -4-[3-(triethyl ammonium sulfonato) phenyl amino] phenylene hydrochloride

Compound C3, Bis -4 [1, 2 diamino phenyl]-phenyl sulfone) hydrochloride

Compound C4, 1, 4 [1, 2 diamino phenyl] Phenylene hydrochloride

THE GRAPHICS ON THIS PAGE WERE NOT AVAILABLE ON THE E-MEDIA VERSION OF THIS PAPER.

Figure S-1. The bipolar nature of PVA-g-PAA-BASPAPS membrane which reduces the methanol crossover due to the molecular dissociation. In (a), the imidazolium group is shown to separate out from the PVA moiety of the chain into a separate phase. In (b), the water is shown to complex with the SO and the imidazolium group. As a result of this, interaction in (d) and (e), the water and the methanol is shown to split into anion and cation and the anion permeates through the membrane giving bipolar effect

(106e) Numerical Investigation in Heat Conduction Effect on Heat Exchange for Mini Cross-Flow Heat Exchanger (MCHE)

Hongyue Duan[1], Wei Mu[1], **Xigang Yuan**[1] and Lingai Luo[2], (1)State Key Laboratory of Chemical Engineering, Tianjin University, Tianjin, China, (2)LOCIE Chambery, Savoie University, Savoie, France

In a traditional heat exchanger, the effect of heat transfer within the wall material driven by the temperature gradient in axial direction of the fluid flow is usually very small because the area of the cross section of the tube wall is much smaller than that of the tube for heat exchange. However, in a mini cross-flow heat exchanger(MCHE), where the heat conducting medium is typically a cubic-shaped block of metal and the hot and cold streams are put into heat exchange via two sets of channels in the block, the heat conduction in the wall material in the direction parallel to the fluid flow may have a significant effect on the performance of the heat exchanger.

In this paper, a 2-channel-layer MCHE was simulated and a periodic boundary condition was applied so that the results could be applicable to entire MCHE. The proposed model and the mesh grid size were proved by the good agreement of the simulated overall heat transfer coefficient with that found in the experimental reported in the literature[1]. Then the effect of heat conduction on heat transfer performance in a MCHE was studied numerically. The simulation results suggest that the performance of the MCHE can be improved by increasing the wall thickness between the adjoining channels in the same layer or by decreasing the wall thickness between the adjacent layers. The heat conduction effect can also be weakened by optimizing the material conductivity or by increasing the diameter of the channels. New correlations were developed based on the simulated results to predict the value of the overall heat transfer coefficient as a function of the operating parameter such as Re in the channel and the design parameters of the MCHE such as the diameter of the channel, wall thickness and thermal conductivity.

[1]Luo L A, Fan Y L, Zhang W W*et al.*Integration of constructal distributors to a mini crossflow heat exchanger and their assembly configuration optimization[J]. Chemical Engineering Science, 2007, 62 (13): 3605-3619.

See more of this Session:Spring Meeting Poster Session and Networking Reception
See more of this Group/Topical:Liaison Functions

(106g) The Effects of Out Gassing Through Cylindrical Walls on the Transport Coefficients

Luma. H. Yaseen[1], **Nada B. Nakkash**[2] and M. N. Latif[1], (1)Al-Nahrain University, Baghdad, Iraq, (2)Chemical Engineering, Al-Nahrain University, Baghdad, Iraq

Abstract

This work concerns with studying the influence of blowing on friction factor and mass transfer coefficient when the flow is normal to a tube bank with outgassing porous cylinders. The friction factor and the mass transfer coefficients are measured for a range of Reynold's number and blowing rate and expressed in terms of various dimensionless groups.

The test rig consists of a rectangular duct (20.3 × 11.3 cm), five rows of three porous tubes in staggered arrangements fixed across the duct with 1.5pitches (both longitudinal and transversal). The outside and the inside diameters of tubes were 25.4 and 21.4 mm, respectively.

The friction factor measurements was studied for Reynold's number range of

$3000 < Re < 17{,}000$ and blowing rate up to 4×10^{-3}. The results show that the blowing intothe boundary layer for the cylinders increases the overall friction factor from 10%to 50% depending on the blowing rate. The mass transfer coefficient was measured for Reynold's number range of $5000 < Re < 16{,}000$ and blowing rate of $10^{-5} < m/G_{\infty} < 10^{-3}$.

A new definition of the blowing rate was considered and expressed by the mass transfer driving force. Two successful formulas were obtained to account the outgassing effect on the mass transfer coefficient. The mass transfer coefficient for the outgassing boundary layer was obtained first by determining the non-blowing coefficient and multiplying it with the correction factor .

The first model was modified to account the density variation associated with the boundary layer change:

$0.51\ Re^{-0.5} \rho_s$

$St_m Sc^{2/3} = \text{--------------}\ (\text{------})^{0.4}$

$(1 + \beta_m)^{4.7} \rho_{\infty}$

With the coefficient of determination R2 = 0.93.

The second model depends on the flat plate correction factor:

$Ln(1 + \beta_m)$

$St_m Sc^{2/3} = 0.5\ Re^{-0.5} (\text{---------------})^{9.94}$

β_m

With the coefficient of determination R2 = 0.92.

These two models are found to be in agreement with the previous literature and a reduction of about 80% in the mass transfer coefficient has been obtained due to the outgassing.

(106h) Down the Hubbert Curve

Ian Sutton, Sutton Technical Books, Houston, TX

Down the Hubbert Curve

In March 1956 M. King Hubbert, who was at that time Chief Geologist with the Shell Development Company published a seminal paper with the title "Nuclear Energy and the Fossil Fuels" (Hubbert 1956). In this lengthy and authoritative paper, which was presented before the American Petroleum Institute in San Antonio, Texas, Dr. Hubbert showed how key fossil fuel resources in the United States, particularly coal and oil, were of finite size and that their output would reach a maximum, and then decline. Dr. King's forecasts for the production of oil in the United States were accurate; he predicted that oil production would peak in the early 1970s, and that is what has happened. (The widely-used term "Peak Oil" is frequently accredited to Dr. Colin Campbell (2001)).

In the fifty years since the publication of Dr. King's paper great advances have been made in finding and extracting oil from what were considered to be marginal fields. But Dr. King's overall conclusion is still valid. Production of oil in the United States remains below the 1970's peak. Although Dr. Hubbert's analysis was confined to the United States, subsequent analysts applied his insights to other parts of the world. For example, the book *Twilight in the Desert: The Coming Saudi Oil Shock and the World Economy* (Simmons 2005) by the late Matthew Simmons achieved wide recognition by showing that the concept of peak oil most likely applies to reserves in Saudi Arabia, and other nations such as Iran and Kuwait.

The purpose of this paper is to examine how oil production and consumption have changed in the last five years, and to examine whether or not the concept of Peak Oil remains valid. Key to this discussion is an understanding of the concept of Energy Returned on Energy Invested (EROEI). It is concluded that the concept of Peak Oil, as first postulated by Dr. King, remains valid and current, and the production of oil peaked in the period 2005-2011 and is now set to move down the right hand side of the Hubbert Curve.

The title of this session is *Challenges in Instituting Energy Savings Project in the Present Market Conditions*. If the production of oil and has indeed already passed a peak then the challenge is not to do with "instituting" energy savings; those savings are going to occur whether we like it or not. The challenge will be to determine how the process industries can restructure themselves given that the energy resources available (both as direct energy and as petrochemical feedstock) will be in decline and that there will be intense demand for those same resources from other areas of the economy.

(106j) Statistical Decision Making for In-Service Inspection of Underground Waste Storage Tanks

Emily Mitchell, Emory University, Atlanta, GA and **Stephen P. Harris**, Savannah River National Laboratory, Aiken, SC

Pitting is a localized form of corrosion that occurs on passivated metal surfaces exposed to aggressive chemical environments. The pitting process involves the breakdown of the passive oxide film and subsequent anodic dissolution of the metal at a local site, while the remainder of the surface remains protected by the film.

Deep pits may result in lowering the maximum level of waste that made be stored in the tank and eventually lead to loss of confinement. Pit depths, established by ultrasonic testing, are subject to variables that cause uncertainty, including the repeatability of metric location, time between measurements, different transducers, and different inspectors.

The generalized extreme value (GEV) distribution is fit to the data gathered from ultrasonic testing and statistical projections are made for the full tank wall interior. A comparison is made with results from other well known statistical distributions. The estimated parameters are used in statistical tolerance limits for bounding pit depth in a waste tank.

The tolerance limits give an indication of extreme values of pit depth for the current condition of the tank walls. The upper tolerance limit for pit depth will then be compared to action levels for pitting. If the upper tolerance limit for pit depth exceeds the action level, more frequent examinations or increased tank surface area inspections are conducted.

(106k) Thermal Inbreathing Requirements of Low Pressure Storage Tanks At Elevated Temperatures

Donald E. Brooks, Pressure Relief Analysis, Lloyd's Register Energy Americas, Inc., Houston, TX

ISO 28300 (API 2000) "Petroleum, petrochemical and natural gas industries - Venting of atmospheric and low pressure storage tanks" provides guidance on determining the required thermal inbreathing flow rates to protect atmospheric and low pressure storage tanks against vacuum due to weather changes (e.g. a decrease in atmospheric temperature)." For heated, un-insulated tanks with vapor space temperatures above 58.9 °C (120 "F), ISO 28300 recommends an engineering review of thermal inbreathing requirements.

This paper describes one approach to evaluating thermal inbreathing requirements for un-insulated storage tanks with vapor space temperatures exceeding 58.8 °C (120 "F)." The tank is assumed to be exposed to the cooling effects of a sudden rain. A simplified unsteady state heat transfer analysis is performed to determine the tank wall and vapor space temperatures versus time. Initial gas mass and pressure are assumed to remain constant in the tank. Gas volume in the tank is calculated as a function of vapor space temperature using the ideal gas law. The rate of volume change is calculated for each time increment. The rate of volume change goes through a maximum value. This maximum value is selected as the thermal inbreathing requirement of the tank.

The calculation methodology described in this paper can be programmed in a spreadsheet for ease of use.

Dynamic Operational Risk Assessment with Bayesian Networks

Shubharthi Barua, Xiaodan Gao, M. Sam Mannan
Mary Kay O'Connor Process Safety Center
Texas A&M University, Jack E. Brown Engineering Bldg., 3122 TAMU, Room 244, College Station, TX 77843
shubharthi.barua@tamu.edu, xiaodan_gao08@mail.che.tamu.edu, mannan@tamu.edu

Extended Abstract Prepared for the Poster Presentation at the
8th Global Congress on Process Safety
Houston, TX
April 1 – 4, 2012

UNPUBLISHED

8TH GLOBAL CONGRESS ON PROCESS SAFETY

Dynamic Operational Risk Assessment with Bayesian Networks

Shubharthi Barua, Xiaodan Gao, M. Sam Mannan

Mary Kay O'Connor Process Safety Center
Texas A&M University, Jack E. Brown Engineering Bldg., 3122 TAMU Room 244, College Station, TX-77843
shubharthi.barua@tamu.edu, xiaodan_gao08@mail.che.tamu.edu, mannan@tamu.edu

Keywords: risk assessment, Bayesian network, dynamic Bayesian network, fault tree, event tree

Abstract

Oil/gas and petrochemical plants are complicated and dynamic in nature. Dynamic characteristics include ageing of equipment/components, season changes, stochastic processes, operator response times, inspection and testing time intervals, and timing of safety system operations, all of which are time dependent criteria that can influence dynamic processes. Conventional risk assessment methodologies can quantify dynamic changes in processes with limited capacity. Several methods are proposed for dynamic risk assessment such as C-DET, D-DET, DORA, etc., but all of them have deficiencies of some kind. Bayesian networks, a graphical inference technique, have some advantageous characteristics for risk analysis over other techniques. This research focuses on providing a framework for dynamic risk analysis based on Bayesian networks. A case study will be provided to illustrate this model.

1.0 Introduction

Oil/gas, chemical, and petrochemical process plants consist of numerous equipment and unit operations, thousands of control loops, and exhibit dynamic behavior. Different hazards and several types of risk are associated with these process facilities. For example, large process areas have the potential to stock a large amount of hazardous chemicals, which can result in the process area being highly congested. It is reported that process risk, i.e., risk in process facilities due to fires and explosions, greatly contributes to the total risk of offshore installations. Also, reactivity, toxic, and mechanical hazards have a significant effect on the total risk of these facilities. Therefore, it is very important to identify hazards, perform risk assessments, and take proper initiatives to minimize/remove hazards and risks; else a catastrophic accident may result.

From case histories, it has been observed that catastrophic accidents have a significant effect on people, environment, and society. Catastrophic accidents such as the Flixborough disaster, the Bhopal incident, and the Piper Alpha disaster caused fatalities and unbearable economic loss. In the past few years, several disastrous accidents in refineries, power plants, and offshore platforms have significantly affected people's perception, and consequently raised concern to emphasize process safety. A series of explosions in 2005 resulting from insufficient process safety and lack of risk reduction measures at the BP Texas City refinery accident resulted in 15 fatalities, 180 injuries, and $1.5 billion in losses [1]. The Kleen Energy power plant explosion in

Connecticut in 2010 is considered to be one of the worst industrial explosions in the USA [2]. In 2010, a fire and explosion resulting from a blowout at the Macondo well resulted in 11 deaths and 17 injuries. Also the continuous spill from the wellhead for 87 days had disastrous effects on the environment and wildlife surrounding the Gulf of Mexico. It has been reported that most of the accidents could be prevented or their severity could have been minimized by effective risk assessment and process safety management. But existing risk assessment methodologies cannot properly quantify various time-dependent effects. Therefore, the objective of this research is to develop a risk assessment methodology with the capability to capture variations due to different time-dependent effects.

2.0 Literature Review

N. Siu [3] summarized available methodologies for risk assessment. For example, the fault tree and event tree methods are two most common methodologies for risk assessment. But both techniques have limited ability to address different time-dependent effects such as equipment or component ageing, stochastic processes, operator response times, and inspection and testing time intervals. N. Siu [3] also described other methodology limitations while applying dynamic system analyses. Therefore, for dynamic systems, conventional techniques are not able to provide accurate results because of their inadequate ability to describe the variation of operational risks as time-dependent deviations, or the changes occurring in the process. Continuous event trees [4], discrete event trees [5], and Markov theory [6] methods have been proposed for dynamic reliability and safety analysis. But all of the proposed techniques have some deficiencies. For example, the Markov theory based model does not consider the effect of inspection on system-state transitions. X. Yang et al. [1] proposed the Dynamic Operational Risk Assessment (DORA) technique, which is able to take inspection and testing time intervals and repair time into account for evaluating operational failure.

Recently, Bayesian networks are receiving wide acceptance in reliability assessment, fault diagnosis, safety, risk, and maintenance analysis. Several authors have [7,8] compared some commonly used methods for accident analysis with Bayesian networks, and described several advantages of Bayesian networks over these methods. N. Khakzad et al. [2] illustrated parallelism between Bayesian networks and fault trees, and discussed the potential of Bayesian networks as a tool for process safety and accident analysis. H. Pasman and W. Rogers [9] described an approach to incorporate Bayesian networks into layer of protection analysis to make it more effective. These studies demonstrate the potential advantages of Bayesian networks application in risk assessment. This research provides a methodology for dynamic risk assessment based on Bayesian networks.

3.0 Methodology and Expected Result

3.01 Bayesian Networks

A Bayesian network, which is a probabilistic graphical model, describes causal influence relations among variables. A set of random variables and their conditional dependencies are represented by Bayesian networks via a directed acyclic graph (DAG). It has the ability to represent dependency among events clearly, accommodate multi mode and continuous random variables, and incorporate information i.e. generic, system specific and expert judgment to support optimum decision making.

A simple Bayesian network is shown in Figure 1. In a binary network, variables are represented by nodes and causal relationships among different nodes are shown by an arc. Conditional probability tables or defined probabilistic relations among nodes represent how one variable is linked with one or many variables. The nodes that influence another variable and have unconditional probability are called parent or root nodes. Nodes that are conditionally dependent on their direct parents are called intermediate nodes. The top node is defined as a leaf node, which is similar to the top event of a fault tree and initiating event for an event tree.

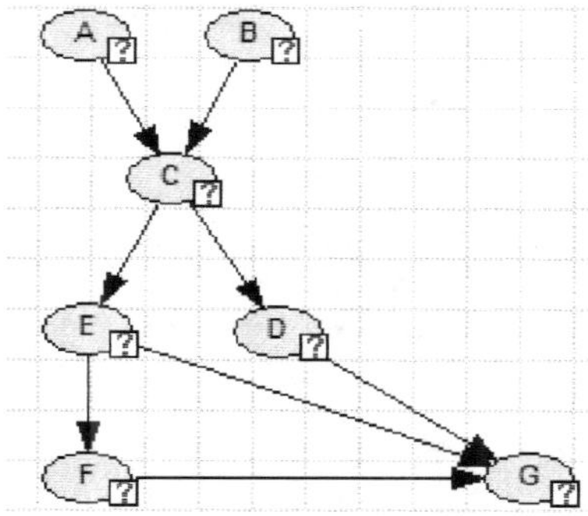

Figure 1: A simple Bayesian network

Let $N = (G, P)$ be a Bayesian network, where

$G = (V, E)$ is a directed acyclic graph; V (random variables) represents nodes; and E represents edges between pairs of nodes of DAG.

P represents probability distribution over V and $V = \{X_1, X_2, \ldots, X_n\}$ is either discrete or continuous random variables [10]. These random variables are assigned to the nodes and the edges.

Bayesian networks can be represented by the joint probability distribution P(V);

$$P(V) = \sum_{X \in V} P\{X|pa(X)\} = P(X_1, X_2, \ldots, X_N) = \sum_{i=1}^{n} P\{X_i|pa(X_i)\}$$

Here $pa(X_i)$ = parent nodes of X_i.

A main advantage of the application of Bayesian networks in risk analysis is the ability to update prior data using Bayes' theorem by incorporating new information as demonstrated by N. Khakzad et al. [2]. Also, Bayesian networks have an advantage of handling different types of uncertainty.

3.02 Dynamic Bayesian Networks

A general Bayesian network is a static model, i.e., the joint probability distribution is usually a representation of a fixed point or an interval of time [11]. A dynamic Bayesian network is a Bayesian network that has the ability to describe the evolution of joint probability distribution over time. Dean and Kanazawa [12] proposed modeling time in a discrete manner to represent the progression of time in a dynamic Bayesian network. In a dynamic Bayesian network, arcs links node/nodes from previous time slice to the node/nodes of the next time slice to represent temporal dependencies among them.

3.03 Methodology

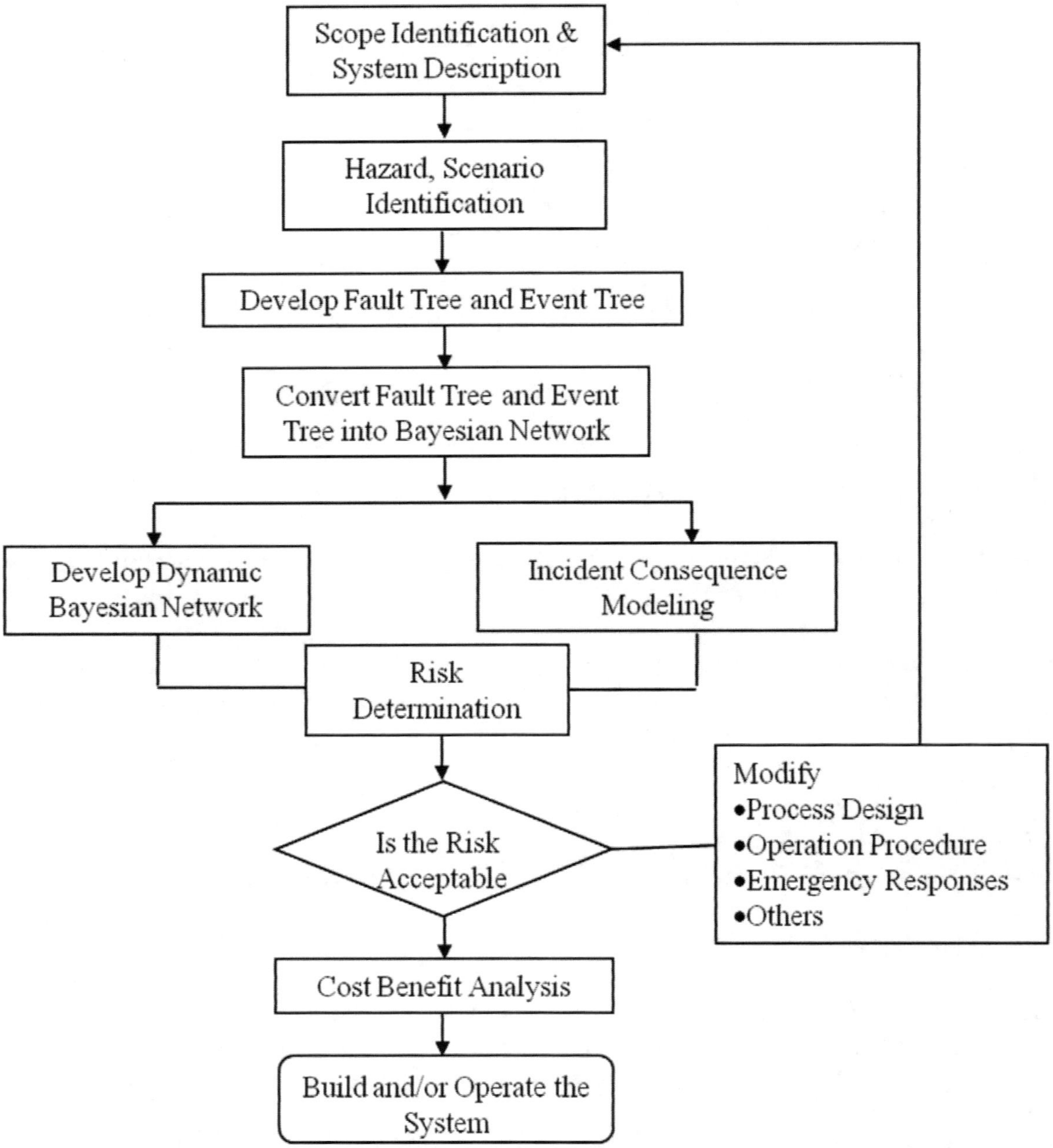

Figure 2: Framework for the dynamic Bayesian network based risk assessment

The objective of this research is to propose a dynamic risk assessment method based on dynamic Bayesian networks. According to the framework, a fault tree and event tree are developed for a system. The objective for following step is to convert the developed fault and event tree into a Bayesian network. Bobbio et al. [13] provided a methodology for conversion of fault trees into Bayesian networks. N. Khakzad et al. [2] showed equivalency between fault trees and Bayesian networks, and described the advantages of Bayesian networks for process safety analysis. Bearfield and March [14] proposed another methodology for generalizing event trees into Bayesian networks. The method has the ability to generalize simple event trees into Bayesian networks in a relatively easy manner, but the method becomes complicated for large event trees. Therefore, in the first phase, a methodology is proposed to convert fault trees and event trees into a static Bayesian network. In the second phase, a dynamic Bayesian network is developed from the static Bayesian network. The dynamic Bayesian network provides probability of a system at different time-slices for different states of the system. The probabilities are then combined with consequence modeling levels to calculate risk. GeNIe, which was developed at the Decision Systems Laboratory at the University of Pittsburgh, is used to create the model [15].

4.0 Conclusions

This article briefly describes a framework based on Bayesian network for dynamic risk assessment for the oil/gas, chemical, and petrochemical industries. This research illustrates the capability of Bayesian networks to represent all component interactions in complex systems, and to update all prior system probabilities in the presence of new information. The advantages of modeling flexibility and efficient computational mechanisms using dynamic Bayesian networks are also discussed.

4. References

[1] Yang, X. and Mannan, M.S., "The development and application of dynamic operational risk assessment in oil/gas and chemical process industry", Reliability Engineering and System Safety 95 (2010) 806-815

[2] Khakzad, N., Khan, F. and Amyotte,P., "Safety analysis in process facilities: Comparison of fault tree and Bayesian network approaches", Reliability Engineering and System Safety 96(2011) 925-932

[3] Siu,N. "Risk assessment for dynamic systems: an overview", Reliability Engineering and System Safety 43 (1994) 43-73

[4] Devooght, J. and Smidts,C. "Probabilistic reactor dynamics-I: the theory of continuous event trees", Nuclear Science and Engineering 1992; 111:229-40

[5] Amendola, A. "Accident sequence dynamic simulation versus event trees", Reliability Engineering and System Safety 1988; 22:3-25

[6] Howard, R. "Dynamic probability systems", Vols. I and II, New York, 1971

[7] Sklet, S. "Comparison of some selected methods for accident investigation", Journal of Hazardous Materials 2004; 111:29-37

[8] Nivolianitou, Z.S., Leopoulos, V.N. and Konstantinidou, M., "Comparison of techniques for accident scenario analysis in hazardous systems", Journal of Loss Prevention in the Process Industries 2004; 17"467-75

[9] Pasman, H. and Rogers, W.; "BBN, a tool to make LOPA more effective, QRA more transparent and flexible, and therefore to make safety more definable", 14th Annual Sysposium, Mary Kay O'Connor Process Safety Center, College Station, Texas, October 2011

[10] Donohue, S.K. and Dugan, J.B.; "Modeling the "Good enough to release" decision usin V&V preference structure and Bayesian belief networks", Reliability and Maintainability Symposium, 2003 Annual, 2003

[11] McNaught, K.R. and Zagorecki, A. "Using Dynamic Bayesian networks for prognostic modeling to inform maintenance decision making", IEEM 2009, Hond Kong, January 2010

[12] Dean, T. and Kanazawa, K., "A model for reasoning about persistence and causation", Computational Intelligence 1989; 5:142-150

[13] Bobbio A, Portinale L, Minichino M and Ciancamerla E., "Improving the analysis of dependable systems by mapping fault trees into Bayesian networks", Reliability Engineering and System Safety 71(3): 249-260

[14] Bearfield G and Marsh W, "Generalising event trees using Bayesian networks with a case study of train derailment", Available online at http://www.agenarisk.com/resources/white_papers/Generalising_Event_Trees.pdf , last modified: 23 February 2010, accessed: 7 January 2012

[15] Decision Systems Laboratory, University of Pittsburgh, GeNIe (Graphical Network Interface and SMILE (Structural Modeling, Inference, and Learning Enginer), Version 2.0 software, 2010. Available online at http://genie.sis.pitt.edu

(137b) Comparing the Performance of Compact Heat Exchange Reactor to CSTR for Fischer Tropsch Synthesis

Hani Gadalla[1], Zhijun Jia[1], Steven J. Vallee[1], Burtron H. Davis[2]and Dennis Sparks[2], (1)Chart Energy & Chemicals, La Crosse, WI, (2)Center for Applied Energy Research, University of Kentucky, Lexington, KY

A commercial Fischer Tropsch catalyst was studied in two Compact Heat Exchange Reactors (CHER) with different channel sizes, and in a continuous stir tank reactor (CSTR). The CHERs were constructed by Chart Energy and Chemicals with one reactor having a 5mm channels and the other having a 6mm channels, while maintaining the same catalyst volumeThe catalyst was crushed and sieved retaining particles ranging in size from 75-125μm. Testing was performed at the University of Kentucky Center for Applied Energy Research in Lexington, KY. The CO conversion, selectivity and hydrocarbon productivity were measured for each reactor over a wide range of operating conditions, as a function of temperature, pressure, H2:CO feed ratio, and flow rate. Both reactor types showed excellent temperature control, but the performances varied at different operating conditions. The larger channel CHER had a higher conversion than the 5mm channel, potentially due to slightly higher reactor bed temperature. The two CHERs approached and even exceeded the productivity of the CSTR for liquid hydrocarbons per gram of catalyst. At higher flow rates the methane selectivity increases at a lower rate for either of the CHERs than for the CSTR.

(116a) The Connection: Our Industry and the STEM Initiative

Toni J. Crawford, Process Technology & Engineering, Fluor Corporation, Sugar Land, TX and Ranessa U. Davis, Process Technology & Engineering, Fluor Corporation, Sugar Land, TX

The Science, Math, Engineering & Technology (STEM) initiative has taken national attention by being implemented from the government's National Science Foundation (NSF) identified need. STEM was essentially born in response to an observation, recognizing the overwhelming lack of interest in pursuing the Science, Math, Engineering & Technology professions within the K-12thgrade students in our country.

Although the pursuit of this initiative is considered valuable from a national level; our industry has not moved with the same haste. In saying this, there is an industry perception of the lack of immediate benefit in developing a link between industry and STEM. This report will address the benefits of implementing industry specific programs which connect the STEM initiative with current industry initiatives; including that of industry economic improvement, longevity and adaptability.

(116b) Teaching Life Cycle Sustainability Assessment Via Online Course

Anthony Halog, Forest Bioproducts Research Initiative/School of Forest Resources, University of Maine, Orono, ME and Najet Bichraoui, School of forest resources, University of Maine, Orono, ME

The concept of sustainability cannot be easily understood because we have been educated through a traditional reductionist approach to learning and research. Whereas the notion of sustainability science requires us to transcend disciplinary boundaries by using integrative, interdisciplinary, holistic and system perspective approaches, university curricula have been developed where teaching and learning has been operating within a silo. Unless we are able to think "*Outside/Beyond the box*", as Albert Einstein articulated in resolving our problems, we cannot effectively impart the idea of sustainability to the next generation of scholars, engineers, scientists, economists and policy makers.

The aim of this work at the University of Maine is to integrate life cycle and integrative thinking into university curriculum. It intends to educate students through a course dedicated solely to sustainability science and find out the most effective method to teach the principles of industrial ecology (IE) and life cycle assessment (LCA). A new interdisciplinary course in IE and LCA has been offered to senior undergraduate and graduate students in engineering, forestry, natural science, agriculture, business administration, economics, human ecology and resource management. The developmental perspective of this course is to transcend disciplinary boundaries by accounting the different backgrounds of students as well as their different levels of interests and passion. We found that the best way to teach environmental sustainability is project-based and by encouraging students to make use of comprehensive online course materials in learning sustainability assessment. A*Blackboard 9.1 Online Learning for Life Cycle Sustainability*has been developed to complement students' learning by having direct access to course materials including LCA course handouts, exercises, assignments, readings, journal articles; key publicly available data, inventory and Weblinks; impact factor information; academic software tools and models for LCA implementation; and tutorial manuals for software and modeling tools. Students have direct links to relevant YouTube and personally recorded audios/ videos; sample LCA reports, checklists and templates; and many others.

The primary benefits and impacts on students are: through case studies and exercises, students have been encouraged to think beyond disciplinary boundaries; and to use systems perspective/integrative thinking approaches. This project has demonstrated the increasing need to incorporate life cycle and transdisciplinary thinking across the university curriculum.

(116c) On Instruction of Statics and Dynamics to a Large Class

Kal Renganathan Sharma, Department of Chemical Engineering, Prairie View A & M University, Prairie View, TX and Ramlingam Radhakrishnan, Civil & Environmental Engineering, Prairie View A & M University, Prairie View, TX

This study is based on instruction of Engineering Mechanics to 303 students in 7 semesters at a HBCU, historically black college and university. The statics section and dynamics section are instructed separately and a 4 credit combined statics and dynamics section is given every semester. The course was assessed using ABET criteria 3(a) and 3(e) skils. The course objectives are for the students to: (i) be able to construct free body diagrams; (ii) be able to apply the two laws of mechanical equilibrium, of translation and rotation; (iii) demonstrate the ability to use 10 different supports in equilibrium, problems; (iv) derive effect of distributed loads by using calculus and laws of equilibrium; (v) learn Coulomb's law of friction; (vi) be abe to device simple machines such as wedges, jack screws and belts using principles of friction; (vii) obtain center of gravity, centroid using integral calculus; (viii) obtain volumes and surface areas of objects of rotation; (ix) compute moments of inertia; (x) derive three laws of kinematics governing velocity, acceleration and displacement; (xi) apply to elevator design and projectile motion; (xii) learn Newton's laws of motion and gravitation; (xiii) derive the space flight trajectory in cyliindrical coordinates; (xiv) apply work and energy theorem; (xv) derive and apply three laws of kinematics of rigid bodies.

A correlation between the students final performances in CVEG 2454 Statics and Dynamics course was sought as a function of their performances in the first test (Bucholtz, 2011). The trend was fairly linear for 209 students. There was some scatter at the lower end of the graph. This can be accounted for by absences in the first test. The scatter was much less at the higher end. The scatter was less at the average to above average end.

PowerPoint slides on FBD, free body diagrams increased the attention of lot of the students. The number of sets of alternate EOEs, equations of equilibrium that can be generated can be 10 sets when 3 different moment become significant. $10 = {}^5C_3$. This was for the two dimensional case.

New theory is needed for moment needed to turn a jack-screw at large thread angles and large coefficient of friction. The current theory gives a mathematical "blow-up" and an unrealistic infinite moment.

The history of how equations of motion and calculus were simultaneously launched with the definition of velocity as rate of change of displacement by Newton and Leibnitz was discussed, More students spent more time with the math skills.

The units of the coefficient in the Belville springs problem was found to be incorrect. This increased the awareness of the importance of each and every term in any equation.

Cubic equations were encountered during analysis of force and displacement problems. Vieta's substitution for depressed cubics can be a method of obtaining closed form analytical solution for cubic equations.

In one problem in projectile motion the student was asked to calculate the angle of the water jet directed at a fire. Upon setting up the equations of projectile motion the equations were reduced to one variable described by one equation. This was quadratic.*Two roots*both real and positive resulted. The two angles meant different points where the water jet will encounter the fire: one before the maximum height and the second after the maximum height.

Students were required to sketch the s-t, v-t, a-t and v-s graphs. The equations to a hyperbola, circle, parabola, ellipse etc were not shown in the Appendix. These were available on the Internet throught wikepedia.org. Once these equatiosn were refered to more students got on board.

In one problem on curvilinear motion, the distance travelled by the car was given as 100 m. The (x,y) coordinates had to be solved for. This needed an lengthier algebraic solution. ∫dL had to be used as shown in chapter 9..0 on the bent rod problem. The integration also involved several methods such as trigonometric substitution, integration by parts and partial fraction expansion. The purport of the normal acceleration calculation using the radius of curvature did not come through clearly. An asymptotic limit can be used, i.e,, the radius of curvature at (100,0) can be assumed for the radius of curvature at (x,y). This leaves the focus of the exercise on radius of curvature and estimate of normal acceleration.

In another problem at the end of the chapter in the textbook on work and energy theorem on nested springs, the spring compressions of the two springs were positive for one and negative for another or vice-versa. Two nested springs with both in compression for absorbing the energy from a falling block is conceivable. It is hard to imagine one of the nested spring in compression and the other in stretched configuration.

Theory behind the formulae used in projectile motion in a x-y coordiantye system was not explained in detail in the book. This was done in detail in the blackboard. The students learnt the significance of v^2. Although the force of gravity changes in direction from negative to positive during the projectile motion the same mathematical expression was shown to be valid for all the phases of the motion in a chosen cartesisn coordinate system.

Power Flight trajectory of a satellite was derived from first principles. The cylindrical coordinates conversion from Cartesian coordinates was shown. This motivated more students. Use of phase angle eased the interpretation of the integration constants.

Many book mistakes were found . This got the students' morale up. For example in a problem in Chapter 13.0 the mass of the train was not needed to complete the analysis of the problem.

Solutions to quadratic equation with two roots both positive could not be interpreted in the spring-mass inclined plane problem. The second root has to be negative with a compressed position interpretation and the positive root the stretched position interpretation.

In one problem, a car was driving up the incline under rain. The stoppage distance under wet and dry conditions were required to be calculated. In one case the stoppage distance was the solution to a quadratic equation. This lead to two positive roots. It was not clear which root is the correct answer. The work and energy theorem was needed to resolve the extraneous root.

The word "erratic motion" can be changed to "domain restricted" solutions in rectilinear kinematics: erratic motion (Chapter 12.0).

Calculation of bank angle for race car track motivated the students further.

In one problem Ball A was released from a tall building and Ball B was thrown vertically upward. The students were asked to calculate the height from the ground and time at which Ball A & Ball B pass each other. It turns out for the numbers given in their textbook, the meeting takes place only when the Ball B descends again. Both A and B move vertically downward! This can be misleading to the novice. The numbers have to be changed so that the Ball A and Ball B cross each other.

Industrial visits to NASA and planetariums are planned.

Students were encouraged to derive equations that they apply to real situations themselves rather referring to formulae derived and kept in books. This improved the fundamental grasp of the course.

The role played by Sir Isaac Newton in the history of the subject was discussed. The story of Galileo was mentioned. How Kepler's equations were integrated into Newton's laws by use of cylindrical coordinates got more students tend to the subject. The planetary laws were around before Newton observed the fall of apple from the tree.

References

REQUIRED TEXTBOOK:

Engineering Mechanics - Statics and Dynamics, R. C. Hibbeler, **Pearson Prentice Hall,** 12th Edition, (2010), Upper Saddle River, NJ. ISBN-13: 978-0138149291

REFERENCES, SOURCES AND FURTHER READING

Philosophiæ Naturalis Principia Mathematica, Sir Isaac Newton, Vol I-III, (1687), Latin, English Translation by R. Bentley.

Engineering Mechanics - Statics and Dynamics, I. H. Shames, **Prentice Hall,** 4th **Edition,** (1997), Upper Saddle River, NJ. ISBN 0133569241.

Engineering Mechanics Vol. I-Statics, J. L. Meriam and L. G. Kraige, **John Wiley and Sons,** (1986), New York, NY.

Engineering Mechanics Vol II – Dynamics, J. L. Meriam and L. G. Kraige, **John Wiley and Sons,** (1986), New York, NY.

Schaum's Outline of Lagrangian Dynamics, Wells, **McGraw Hill Professional,** (1967), New York, NY.

Vector Mechanics for Engineers, F. P. Beer and E. R. Johnston, **McGraw Hill Professional,** 8th Edition, (2007), New York, NY.

Engineering Mechanics : Dynamics, A. Pytel and J. Kiusalaas, **CENGAGE Learning,** Third Edition, (2010), Stamford, CT.

Engineering Mechanics: Statics and Dynamics, F. Constanzo, M. E. Plesha and G. L. Gray, **McGraw Hill Professional,** (2010), New York, NY.

(116d) Trends of Chemical Engineering Curriculum in Iran and New Challenges

Yousef Jalali, University of Minnesota, Minneapolis, MN

Iran is an ancient country and one of the cradles of knowledge. These days, the educational system in Iran encounters major challenges that influence chemical engineering education as well as other disciplines. We can observe the decline of the value of knowledge in the recent years. A revolutionary change in the authority's viewpoints would be required to overcome the educational problems. In general, the history of higher education in Iran reflected the sociopolitical events. Tribal economic system, anarchy, rebellion, absolute authority, violence, foreign interference, contradiction between traditionalism and modernism and an impatient political culture were major obstacles in the forming of educational institutions. In the beginning of the 20th century after relative stability and in response to a modernist program, the university was formed by the government. Following the Islamic Revolution in 1979 and in the period called the Cultural Revolution; universities were closed for three years for desecularization and Islamization. Today, despite a wealth of young talent in Iran, students do not have the ability to satisfy goals and meet standard international expectations due to lack of policy. Undoubtedly most of our educational problem has roots in the authority's policy. However, major shortcomings in curriculum structure could be modified and help students tackle new technologies and demands. Chemical engineering education in Iran has undergone two major paradigms unit operations and engineering science. However the content of curriculum has not been change in response to demands on new technologies and professional skills that influenced the curriculum siginificantly in the last few decades. In this study, a brief history of chemical engineering education in Iran is presented. The study then suggests improvements that could be made to improve the undergraduate chemical engineering program in Iran considering the industrial and societal demands.

Teaching Chemical Engineering to Non-Chemical Engineers

Extended Abstract

For the past twenty years, AIChE has offered a public course several times a year which is basically an overview of chemical engineering basics, designed for operating technicians, chemists, other engineers, and managers of chemical engineers who may have no background in chemical engineering. This course has also been attended by Department of Homeland Security chemical plant inspectors, CCPS personnel, and EPA attorneys. This course has no calculations or design work, but focuses on general concepts and relationships between variables such as Reynolds and Prandtl numbers. In 2003, this course received a significant upgrade to include substantial content in the areas of safety and reactive chemicals, special unit operations such as chromatography, and solids handling. A series of problems, requiring thinking about process and unit operations' response to changing conditions was also developed. Over the past ten years, this course has also been delivered on site to companies and organizations including Pfizer, Department of Homeland Security, Victoria (TX) community college, Gelita, Solutia, and FMC.

Over this period of time, these are the major learnings and focus areas of most importance and focus:

1. Scale down vs. scale up. What is meant here is to get the attendees to think about the operating large-scale process and conceptually shrinking it down vs. scaling up the mini-plant or laboratory model of a process. These two thought processes do not end up at the same place. A larger and larger stirred flask or reactor does not necessarily mean that a large stirred reactor is what is needed in the full scale plant. If the kinetics and reaction rate information is well understood, multiple reactor design options, including continuous flow reactors may be the best choice for the large scale plant.
2. Reaction engineering fundamentals. It is emphasized that a simple "rate vs. time" model is not reaction engineering kinetics and is not sufficient to optimally design a chemical reactor. The exponential nature of kinetic rate curves with temperature is emphasized.
3. Reaction rates vs. heat transfer. As part of the emphasis on safety and reactive chemicals, a strong emphasis on the difference between the logarithmic nature of reaction kinetics and the linear nature of heat transfer rates is emphasized with a real case study.
4. Endothermic vs. exothermic reactions. This basic difference is reviewed and the concept of chemical equilibrium associated with sulfuric acid manufacture is reviewed.
5. Solids handling. The addition of material related to solids handling focuses on development of material properties of solids at the point of use. Angle of repose and tapped/bulk density concepts are reviewed with examples from real situations. How solids are tested is also reviewed.
6. Basics of process control are reviewed and the need to balance control response with the reaction kinetics of the system are covered.

7. Basic concepts of distillation, relative volatility, and azeotropes are also covered with case studies.
8. Basic concepts of fluid flow and pump characteristics are reviewed including Reynolds number, NPSH, and

Videos from the CD-ROM produced by the AIChE Foundation are used to illustrate many of these concepts.

The areas of most impact on attendees is the general area of physical property data (or lack thereof!) on proper design of fluid handling systems, mass transfer unit operations, and solids handling systems.

(140a) "Reverse Circular Teaching": A Pragmatic and Effective Way to Increase Successful Understanding

Phaneendra B. Kondapi, Flow Management, FMC Technologies, Houston, TX; Mechanical Engineering, University of Houston, Houston, TX

The Reverse Circular Teaching method is about making the student fully understand the basic concepts of the course and motivate towards successful learning. Effective teaching by the instructor and satisfactory learning by the student can be achieved by normal teaching methods but in order to make the student fully understand the concepts of a real world problem, a new method of teaching is necessary.

In this competitive world, industry is looking for people who are not just bright and smart but also be able to get into productive work right away the moment they are hired. So the need for practical knowledge right at the college level is very much increasing. Design and teach a course to meet the expectations of both industry as well as academia with more of practical content while balancing the theoretical part of it is challenging.

In order to meet that challenge a new effective and practical teaching method has been developed to teach a course at University of Houston, probably the first three credit-hour university course in the US or may be in the world that has been designed to meet direct industrial applications with academic standards. The method is about solving a practical design problem not after learning the theory but learning the theory after understanding the problem.

(140b) Liquid Level Systems for Process Control Education and Researches

Jietae Lee, Department of Chemical Engineering, Kyungpook National University, Daegu, South Korea and Dong Hyun Kim, Department of Chemical Engineering, Kyungpook National University, Taegu, South Korea

Process control is closely related with static and dynamic characteristics of processes. Experiments will improve the process control course. For this, experimental systems that are small and do not produce wastes are required so that they can be used in the class room. Several such systems are available for areas of mechanical engineering and electrical engineering. In the process control area, liquid level systems are adequate. However, most of liquid level systems available are too simple to illustrate various process static and dynamic characteristics (Astrom, K.J. and Ostberg, A.B., "A Teaching Laboratory for Process Control," IEEE Control Systems Magazine, 6, 37-42, 1986). Recently Johanson (Johansson, K.H. "The Quadruple-Tank Process: A Multivariable Laboratory Process with an Adjustable Zero," IEEE Trans. Control Systems Technology, 8, 456-465, 2000) proposed a liquid level system with four tanks, four manual valves, two actuators and two level sensors. This system can be made to have an unstable zero and is found to be very useful for process control education and researches. Several research papers have also been available (for example, Mercangoz, M. and Doyle III, F.J., "Distributed Model Predictive Control of a Four-Tank System," J. Process Control, 17, 297-308, 2007).

In this work, a novel liquid level system which consists of changeable two tanks, one solenoid valve for the actuator and one level sensor is proposed. It is smaller and simpler than the previous ones. Most of process characteristics which are in process control textbook (Seborg, D.E., Edgar, T.F. and Mellichamp, D.A., Process Dynamics and Control, 2nded. Wiley, N.Y., 2004) are provided. They include

l Linear Element with 1stOrder, 2ndOrder, and so on.

l Integral Element

l Time Delay Element

l Inverse Response Element (Unstable Zero Element)

l S-shape Nonlinear Element (Mimic pH Process)

l Nonlinear Element having Input Multiplicity

l Nonlinear Element having Hysteresis (Mimic Ignition of Exothermic Reactor)

l Nonlinear Element with Limit Cycle Instability

USING PFD DIAGRAMS AS TEACHING TOOL FOR NON CHEMICAL ENGINEERS

Gaitán Lady[1], García Johanna[1], Martin Luisa[1], Camargo Gabriel Environmental Eng. Dpt[1] and Mechanical Engineering Dpt. Universidad Libre Colombia

Introduction

Teaching process or design process for no Chemicals engineer can be improve means flow diagrams. Also important concepts learned in courses such as thermodynamics, chemistry or fluid mechanic can be reinforced by using flow diagrams.

°This paper is about the implementation of flow diagrams as learning tools in a material balance course for environmental engineering at Universidad Libre in Bogota, Colombia, The concepts and methodology of materials balance will be used and applied in other areas (2) (water treatment, air pollution soil analysis etc.).

The syllabus of materials balance begin with, Definitions, separation process and different kinds of separation process,(1) then follow with Unit and Dimension, general equation for material balance with and without chemical reactions and application exercises. All these topics are later reinforced with the concepts of flow diagrams: process equipment, auxiliary equipment, utilities and control systems.

For the implementation of this method, as a pedagogical tool, the students are interrogated (something similar a survey) about the meaning of words like raw materials, products, properties of substances, utilities, differences between pump and compressor, among other- After that, the concepts of flow diagrams and kinds of flow diagrams (Block flow diagram BDF, Process Flow Diagrams PFD and Piping and instrumentation diagrams P&I D), are given. The activities end with homework of identification of raw materials, products and utilities for a specific process. The goals are several: identify the utilities and their possible impact in the environment looking for sensible and latent heat in the process and different streams process, etc.

The final part of the material balance course is a design project of a chemical product and the draw of the different flow diagrams with a preliminary budget and material properties evaluation and environmental emissions.

This methods has been successful in the environmental Engineering department because integrate other courses (thermodynamics, chemistry, water treatment, etc). And actually this method is been tested in Industrial Engineering Dept. The guide used for the diagrams explanation and application is Turton's book *Analysis, Synthesis, and Design of Chemical Processes* (3)

The impact of this methodology in the course was evaluated by the student´s evaluation, and the report of the final projects

Application Example

Using the link *http://www.che.cemr.wvu.edu/publications/projects/index.php*, (3) some data registered in this place are taken for reinforced concept related with concentration, viscosity relation between molar flow and mass flow. The dates are changed and the students are motivated to find the errors present in the tables. or figures.

For example the data tables 1 are used to formulate questions like these:

Table 1. Data used for question related with thermodynamics

UTILITY	lps	cw	rw
Equipment	E -405	E-406	E- 407
Temperature In (ºC)	100	30	5
Temperature Out (ºC)	100	40	15
Flow (tonne/h)	1,28	77,8	13,6
Low Pressure is supplied at 6 bar (sat) from the header			

Is some mistake whit the data in the equipment E405?
Why is the reason for temperature not change? Some trouble with detectors?
What kind of heat is been transferred in this equipment?
What kind of heat is been transferred in the equipments E- 406 and E407?

Similar question are formulated for fluid mechanic concepts such as viscosity and pumps and compressor.

References

Use "References" as a heading name, with your listing that follows in the base font size (12pt). Number each reference consecutively (1, 2, 3, etc.) and single space between each reference. Indent after each reference number. Do NOT use end notes. Refer to the example below.

1. King Judson (1980), "Separations Processes" New York NY Mc Graw Hill pp. 1-50.
2. Allen D. and Shonnar D "Green Engineering: Environmentally Conscious Design of Chemical Processes" (2002 Upper Saddle River) Prentice Hall.

Turton R., Baillie R., WhitingW. And Shaewitz A." *Analysis, Synthesis, and Design of Chemical Processes* "(2009) Boston MA. Prentice Hall.

(140d) A Microfluidic Platform for University-Level Analytical Chemistry Laboratories

Jesse Greener[1,2], Ethan Tumarkin[1], Michael Debono[1], Andrew Dicks[1] and Eugenia Kumacheva[1,2],
(1)Department of Chemistry, University of Toronto, Toronto, ON, Canada, (2)FlowJEM Inc., Toronto, ON, Canada

We demonstrate continuous flow acid-base titration reactions as an educational microfluidic platform for undergraduate and graduate analytical chemistry courses. A series of equations were developed for controlling and predicting the results of acid-base neutralization reactions conducted in a microfluidic format, including the combinations of (i) a strong base and a strong acid, (ii) a strong base and a weak acid, and (iii) a strong base and a multiprotic acid. Microfluidic titrations yielded excellent repeatability. The small experimental footprint is advantageous in crowded teaching laboratories, as it offers limited waste and exposure to potentially hazardous acids and bases. This platform will help promote the utilization of microfluidics at an earlier stage of students' careers.

(140e) Using LEGO NXT Robotics, Vernier Sensors and Imagination to Build An Integrated Freshman Engineering and Service Learning Program

Bill B. Elmore, Dave C. Swalm School of Chemical Engineering, Mississippi State University, Mississippi State, MS

Over the past six years, freshmen in the Swalm School of Chemical Engineering at Mississippi State University have "practices" chemical engineering fundamentals and reached out into the K-12 community to engage*future engineers*using systems built from their own imaginations with LEGO NXT™ robotics, Vernier®sensors and small-scale pumps, pipes, tanks, etc. Now a regular feature of our K-12 outreach, these activities are enthusiastically embraced by students—elementary through high school, and are proving to be a relatively-inexpensive means of providing a significant number of students (e.g. 50-60) a simultaneous engineering design experience in small groups of three to four each. Such hands-on activities in a high-energy "studio" environment lead to numerous learning opportunities among our chemical engineering and K-12 students alike. In this paper, we report on several attractive features of this program including: 1) improved techniques for recruiting students to engineering studies, 2) stronger "visualization" of engineering concepts among our engineering freshmen and 3) a stronger sense of the need for life-long learning and community service among our engineering undergraduates.

AUTHOR INDEX

American Institute of Chemical Engineers
3 Park Avenue
New York, New York 10016-5991

CURRAN ASSOCIATES INC
proceedings.com